Edward Bellamy

The Student's Guide to Surgical Anatomy

Edward Bellamy

The Student's Guide to Surgical Anatomy

ISBN/EAN: 9783337365622

Printed in Europe, USA, Canada, Australia, Japan

Cover: Foto ©berggeist007 / pixelio.de

More available books at **www.hansebooks.com**

THE

STUDENT'S GUIDE

TO

SURGICAL ANATOMY:

BEING A DESCRIPTION OF THE MOST IMPORTANT SURGICAL
REGIONS OF THE HUMAN BODY, AND INTENDED
AS AN INTRODUCTION TO

OPERATIVE SURGERY.

BY

EDWARD BELLAMY, F.R.C.S.,

ASSOCIATE OF KING'S COLLEGE, LONDON; SENIOR ASSISTANT-SURGEON TO THE
CHARING CROSS HOSPITAL; SURGEON TO THE ROYAL INFIRMARY FOR
CHILDREN, WATERLOO ROAD; AND TEACHER OF OPERATIVE
SURGERY IN THE MEDICAL SCHOOL OF CHARING
CROSS HOSPITAL.

WITH ILLUSTRATIONS.

PHILADELPHIA:

HENRY C. LEA.

1874.

PREFACE.

A CONSIDERABLE experience as a teacher has convinced me that there is room for a small work on what might be termed Applied Anatomy. With the exception of my friend Mr. Galton's translation of Professor Roser's work, there is no English handbook of the kind within the reach of the generality of students.

Pupils are apt to lay aside their anatomical studies after having passed their primary examination, and to be confused at finding a considerable amount of Surgical Anatomy required of them when they present themselves for their final or pass.

With a view of assisting them, this work has been prepared. A knowledge of Descriptive Anatomy is presupposed, and such regions of the body as do not seem to bear directly upon the operative or more practical parts of Surgery, have been either merely referred to or entirely omitted.

I do not hesitate to state that I have in one or two instances availed myself of methods of arrangement adopted by others; yet all the statements have been confirmed by actual demonstration, and may be regarded as an embodiment of the remarks made to students attending my course of Operative Surgery at Charing Cross Hospital.

The engravings are, in some instances, borrowed by permission from Mr. Heath's excellent work on Practical Anatomy, or have been drawn upon wood by myself, Mr. Wesley, and Mr. Sherwin, from nature, or have been adapted from such sources as will be found acknowledged in their proper places.

I have to express my obligation to my friend Dr. James Cantlie, M. A., M. C., Demonstrator of Anatomy at Charing Cross Hospital, for his kindness in revising the proofs, and for other valuable assistance.

EDWARD BELLAMY.

MARGARET STREET, CAVENDISH SQUARE,
October, 1873.

CONTENTS.

ERRATA.

Page 82, line 10 from top, the sentence should read thus: "The pulsations of the vessel are in reality felt beneath its anterior border," &c.

Page 89, first line, *for* "superjacent" *read* "subjacent."

Page 90, 12 lines from bottom, *for* "process" *read* "artery."

LIST OF ILLUSTRATIONS.

B

INTRODUCTION.

As the subject of Regional Surgery and Surgical Anatomy bears directly upon Operative Surgery, it may not be considered out of place to remind the student of the necessity of making most careful inspection of the body as a whole before he attempts the more minute and detailed examination of its various parts. For this purpose both the living model and the dead subject should be examined together. For such examination the body should be lying down—in fact, in the position a patient would be placed in for a surgical examination or operation. By the side of the body should be placed an entire articulated skeleton. Careful notice is to be taken of all the surface-markings, and of the superficial bearings of all prominent underlying structures, such as the subcutaneous surfaces of the bones, ligaments, tendons, and bursæ; the proper swellings, or *contouring*, of the muscles, both at rest and in action; the course of the superficial and deep vessels; the change of aspect in regions, dependent upon alteration of position; the course and direction of the several natural passages of the body; the anatomical relations of the lines of incision required in the various operations of surgery; and the altered positions of bones when dislocated compared with the normal ones.

In studying Regional Anatomy the parts must be regarded as being wounded, either by the surgeon's knife or by some weapon, or displaced by accident. In the various stages of a dissection made in the prosecution of Descriptive Anatomy, it is often the end and aim of the dissector to make a clean or " pretty " preparation, in following out the different vessels, nerves, &c., and for this purpose it is quite right that all pains be taken, but the student must remember that the more he cleans, the more he destroys the actual relation of the parts as they would be met with in an operation ; and, moreover, he must remember that the very fasciæ he so studiously removes, are of the greatest importance in Surgical Anatomy, and their removal destroys surgical continuity.

The want of material in our schools is the great drawback to the study of Topographical Anatomy, as bodies cannot be spared for such sections and special examinations as a consideration of the various operative proceedings of surgery suggests.

THE STUDENT'S GUIDE

TO

SURGICAL ANATOMY.

CHAPTER I.

SURGICAL ANATOMY OF THE CRANIAL REGION.

Regions.—The regions into which the cranium is divided for the sake of surgical reference, differ in number and extent in the various works upon the subject; thus some make four—viz., the occipito-frontal, the temporal, the auricular, and the mastoid; but for simplicity and more ready reference the following would seem to suffice, viz., the *occipito-frontal* and the *temporo-parietal.* The external investment of the cranium or scalp varies in structure in these several regions, and will be considered in the dissection of each.

The occipito-frontal region is oblong in shape. Its limits are—anteriorly, the anterior margins of the roofs of the orbits, and the articulations of the frontal with the superior maxillary and nasal bones; posteriorly, by the superior curved line of the occipital bone, and on each side by the temporal ridge.

Dissection.—An incision is to be made, commencing

2

in front at the root of the nose, and is to be carried backwards to the occipital tuberosity ; a second at right angles to it across the scalp from ear to ear. Care must be taken that the depth of the first incisions extends no further than the hair-bulbs. The flaps so formed are next to be very carefully reflected forwards and backwards.

The integument is tolerably thick, smooth, and somewhat uneven in surface. It is very rich in sebaceous follicles, and these, by the inspissation of their contents and closure of their excretory ducts, constitute the encysted tumors or *wens* so often met with in this region. It is highly vascular, and is frequently the seat of aneurisms by anastomosis, erectile tumors, &c., and has immediately subjacent a dense lamellated cellular tissue, containing a little nodulated fat and the hair-bulbs, adhering firmly to the underlying tendinous expansion of the occipito-frontalis muscle or epicranial aponeurosis. This aponeurosis is extremely thin over the vertex of the cranium, but very thick in the temporal region, and freely movable. Effusions of blood may take place either above or below this structure in the loose areolar tissue separating it from the pericranium. Thus a " black eye " can be produced by a blow on the back of the head, by the gravitation of the blood downwards and forwards into the loose subcutaneous tissue of the forehead and eyelids. In phlegmonous erysipelas of the scalp the pus burrows under this aponeurosis, so that free incisions down to the bone and counter-openings are necessary. Beneath this cellular layer is the pericranium or external periosteum, which is much stronger in the child than in the adult; it is frequently the seat of periostitis and of nodes.

The arteries supplying this region are—in front, the supra-orbital, the frontal, and the superficial temporal, thus forming a free inosculation between the internal and external carotid arteries. Behind, the occipital and the posterior auricular branches of the external carotid artery freely anastomose with each other, and with the abovementioned vessels.

The superficial temporal artery, which is easily seen beneath the skin in its tortuous course, is liable to injury, which, in the event of the main trunk being divided, either entirely or partially, may be serious by the formation of a false aneurism from the escape of blood beneath the tissues. In such a case the tumor must be opened and the clot turned out, and *both* ends of the bleeding vessel tied. If the operation of arteriotomy be required, the anterior branch of this vessel is the one selected, just at the spot where it begins to be covered by the hairy scalp. In performing this operation a small puncture only is necessary, and when complete the vessel should be *entirely* divided, so that by the retraction of the cut ends the formation of a false aneurism may be prevented. In such plastic operations as the restoration of the upper eyelid, or of rhinoplasty, the preservation of the supraorbital arteries is of great importance for the proper nourishment of the flap.

Veins.—The arteries are generally accompanied by small veins; one vessel in particular, the *vena præparata*, which is situated in the middle of the forehead, and is plainly visible during bodily exertion or under excitement, is worthy of note. This vein has been proposed as suitable for venesection.

The nerves supplying the muscles and integument, are in front, the supra-orbital and supra-trochlear branches of the frontal, and some few inosculating branches of the facial. These nerves are frequently the seat of neuralgia, and for its relief division of the frontal nerve has been proposed at its point of exit from the supra-orbital notch.

In infancy this region is frequently the seat of ceph-alhæmatomata or blood tumors, occurring either from compression of the cranium during parturition, or from a collection of blood beneath the pericranium, which is very loosely attached to the bone at this period.

Structures divided in cutting down upon the Bone in the Occipito-frontal Region.—Skin, subcutaneous cellular tissue, occipito-frontalis muscle, and epicranial aponeu-rosis, a thin layer of lax cellular tissue, and the pericra-nium; anteriorly the vessels divided are the temporal, the frontal, and the supra-orbital, with the supra-orbital, supra-trochlear, and branches of the auriculo-temporal nerves, and posteriorly the occipital vessels and greater and lesser occipital nerves.

The Temporo-parietal Region.—The superior limit of this region is the lateral boundary of the occipito-frontal, and its inferior limit is a line drawn from the external angle of the frontal bone to the mastoid process of the temporal.

Dissection.—*The integument and subcutaneous cellular tissue* are pretty much the same as in the preceding re-gion. The structures met with on reflecting the integu-ment are—a superficial aponeurosis, very tough, upon which is the temporal artery; a second aponeurosis lined with fat and cellular tissue, and the small auricular muscles. *The temporal fascia* is attached above to the curved line limiting the temporal fossa, and to the zygoma below; by its under surface it gives attachment to the temporal muscle, in which lie the deep temporal arteries from the internal maxillary, with their accom-panying veins. The deep temporal arteries freely inosculate with the superficial and with the occipital and posterior auricular which supply the hinder part of this region. The arrangement of the several aponeuroses in this region determines the course taken by purulent collections; thus, supposing the matter to be deepseated, it will make its way into the zygomatic fossa, and when superficial it will be limited by the tough aponeurosis already mentioned. The posterior part of this region, which contains the mastoid process of the temporal bone, is the seat of an operation proposed for the opening of the mastoid cells with the object of giving vent to matter pent up in them, should it not find its way either by the Eustachian tube or through the external auditory meatus. Fractures in this region are frequently complicated with laceration of the middle meningeal artery, which runs over the internal aspect of the cranium.

This region contains numerous lymphatic ganglia, which become remarkably indurated in constitutional syphilis.

The structures divided in cutting down upon the bone in the temporo-parietal region are—the skin, subcutaneous fascia, epicranial aponeurosis, superficial temporal aponeurosis, deep temporal aponeurosis, temporalis muscle and tendon, with the superficial and deep temporal vessels, nerves, and lymphatics.

SURGICAL ANATOMY OF THE CRANIUM.

The structure of the cranial bones forming the vault of the skull consists of three layers; an outer, formed of tough compact tissue; an intermediate, the diploë, soft and spongy, having the diploic veins ramifying in its substance; and an inner, hard and brittle. These diploic veins, after injury followed by suppuration, are liable to inflammation; which circumstance explains the formation of secondary deposits of pus in various parts of the body, most frequently in the lungs and liver. The brittleness of the internal layer is of surgical importance from the fact that, in blows on the head, it is more liable to be fractured than the outer; and cases have occurred where it has been broken without any apparent depression whatever of the external, giving rise to symptoms of compression which would otherwise have been difficult of explanation. The diploë is not easily distinguished in young persons. In the application of the trephine, the varying densities of the layers of the skull must be borne in mind, as the pressure upon the instrument on its first application must be firm and steady, until the external table is perforated, when there is less resistance. When it is quite perforated the blood of the diploë will be seen in its teeth. The pin of the trephine is now to be withdrawn, to avoid its being pushed through the inner table into the dura mater and encephalon; the inner table, though thinner, will be found to offer more resistance to the saw edge. There are certain localities in the skull where the application of the trephine should be avoided. These are—over the longitudinal sinus, the anterior inferior angle of the parietal bone, because of the middle meningeal artery, over the occipital tuberosity, and over the sutures.

The arrangement of the various sutures of the bones forming the vault of the cranium has greater interest for the obstetrician than the surgeon, as any peculiarities connected therewith exist normally only in fœtal life or early childhood. The bones of the skull in the fœtus or newly-born child are flexible, and between their undeveloped sutures are the *fontanelles*, the position of which is of importance. The sagittal suture, extending from the root of the nose to the occiput, is crossed at right angles by the coronal; at the point of intersection of these sutures is the anterior and larger fontanelle. At the point where the lambdoidal suture crosses the sagittal is the posterior fontanelle, generally closed at birth, and recognizable by the peculiar convergence of the three sutures. Hernia cerebri is the result of incomplete closure of these spaces, which, however, are generally ossified by the fourth year.

Fig. 1.

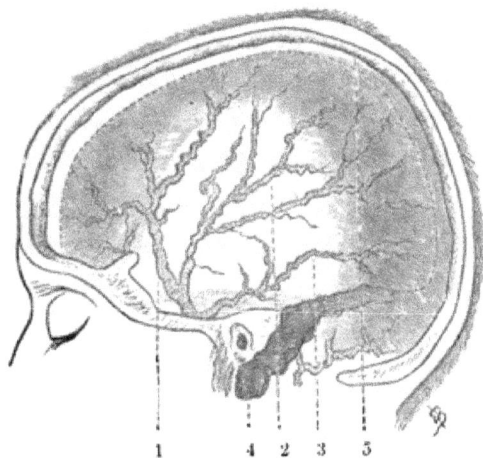

Diagram of structures to be avoided in use of trephine. 1, 2, 3. Branches of middle meningeal artery. 4. Lateral sinus. 5. Superior longitudinal sinus.

Fracture of the base of the skull by "contre-coup" is denied by some, on the ground that the shock is resisted by the cranium, and that the results of such shocks, as in the case of architectural arches, are lost upon its supporting pillars, which, in the frontal region, are the malar and sphenoid bones, in the parietal the temporal bones, and in the occipital region the ribs of the occipital bone itself.

In almost all cases, when the cranium is struck, the parietal region is the seat of the injury; the bone is fractured at the spot, and the line of fracture runs through the temporal bone, which, from the fact of its containing so many cavities and foramina, its texture, and the inclination of the axis of the petrous portion, readily gives way. A fracture of the base may also occur from a fall on the feet or on the buttock, the force being transmitted along the spinal column, and meeting the skull at the condyles. Rupture of the brain-substance, however, is common by *contre-coup*. The course of such fractures of the base may be anatomically determined by the effects produced upon the nerves issuing from the skull, the most frequent being facial paralysis from lesion of the portio dura whilst in the aqueductus Fallopii.

Fractures of the base of the cranium are generally associated with ecchymosis of the eyelids and effusion of blood from the external auditory meatus. The existence of subconjunctival ecchymosis is of great importance in the diagnosis of this injury; the escaped blood infiltrates the cellular tissue of the orbit, passes through the openings in the capsule of Tenon, and so gets into the subconjunctival cellular tissue (*vide* Orbital Region).

The surgical anatomy of the *temporal* bone presents considerable additional points of practical importance, as it contains the organ of hearing and the parts accessory to it.

The *external auditory canal* has a direction inwards and forwards, describing a slight general curve, the concavity of which is downwards. The outer third of the

passage is formed by a tubular prolongation inwards of
the cartilage of the external ear, which, however, is not
complete at the upper part; and its inner two-thirds by
the canal in the temporal bone.   There are several small
fissures in the cartilaginous part, which are sometimes
very wide apart—a circumstance explaining the passage
of pus *into* the meatus, from abscesses which have formed
external to it.   In length the meatus is about an inch
and two or three lines; but owing to the obliquity of the
attachment of the membrana tympani, its anterior wall
is about one-quarter of an inch longer than the posterior.
Its narrowest diameter is about the middle, and in
making an examination with the speculum auris the instrument should not be introduced further than this point.
It must be borne in mind that in young children the
meatus is very shallow, the bony part consisting only of
a small ring of bone, deficient at the upper part, to which
the membrana is attached.

To facilitate the introduction of the speculum, the auricle should be drawn upwards, backwards, and a little
outwards; this renders the canal tolerably straight.

The *membrana tympani* is, on examination, grayish in
color, its fibrous structure looking radiated, slightly conical, with the apex directed inwards, and placed very
obliquely at the bottom of the meatus.   The handle of
the malleus is seen through the membrane, not quite
vertical, but inclining a little backwards.   The points of
practical importance connected with the *tympanum* are
these—that its upper aspect and floor are formed by thin
lamellæ of bone separating it from the cranium and from
the canal for the internal carotid artery, so that disease
of the bone causes death, either by involving the dura
mater and brain, or from ulceration into the vessel.   The

close vicinity of the carotid artery and lateral sinus readily accounts for the escape of blood from the ear in fracture of the base of the cranium.

The *Eustachian tube* is the means of communication between the internal ear and the pharynx, and serves to maintain the balance of air on either side of the membrana tympani.

Its internal orifice is at the anterior internal aspect of the tympanum. The tube is directed downwards and forwards, and terminates in a flattened valve-like opening in the pharynx, just behind and a little above and external to the inferior meatus of the nose. Its mucous membrane is continuous with that of the pharynx. The pharyngeal extremity of the tube is in close relation with the tonsil—a fact which explains its temporary occlusion in enlargement or inflammation of that gland.

(The operation of introducing the Eustachian catheter is explained at p. 31.)

SURGICAL ANATOMY OF THE FACE.

Dissection.—Before commencing a dissection to display the various structures met with in the integument of the face, a little tow should be inserted into the eyelids, buccal cavity, and nostrils, to make tense these regions. An incision is then to be made from above the zygoma to the angle of the jaw, and another meeting it along the base of the jaw to the middle of the chin. The skin-flap is then to be raised from behind forwards, and left adherent along the middle line. Great care must be taken in so doing, as the facial muscles, or " muscles of expression," are inserted into the skin, and are easily removed in dissection.

The skin of the face is remarkably thin, freely supplied with vessels, nerves, and follicles. The subcutaneous cellular tissue is dense, and contains (except on
the eyelids) a good deal of fat. A description of the attachments, relations, and uses of the muscles of the face
will be found in works on general or artistic anatomy.

Arteries.—The chief arterial supply of the face is derived from the external carotid, its facial, internal maxillary, and transverse facial branches, and from the
ophthalmic branches of the internal carotid. Their
inosculation is remarkably free. The normal course of
the facial artery, when it appears on the face, is just anterior to the masseter muscle, where it is subcutaneous,
and here only is it in actual relation with the accompanying facial vein, which is almost straight, and lies to
its outer side. The vessel ascends from this point tor-
tuously, more particularly so in old persons, towards the
corner of the mouth, side of the nose, and inner angle of
orbit, where it inosculates with the ophthalmic. The
artery lies at first under the platysma, and further on in
its course is covered by some thin fibres of the zygomat-
icus major. The chief named branches are the inferior
labial, running between the lower lip and the chin, and
distributed to its integument; the coronary, superior
and inferior, distributed to each lip and to the septum of
the nose; the lateral nasal, to the side of the nose; and
the angular, a large branch going to the inner angle of
the orbit, generally seen pulsationg under the skin. The
transverse facial artery is a branch generally of the temporal, lying by the side of the duct of the parotid gland
(Steno's), accompanied by branches of the facial nerve.
Those branches which the internal carotid supplies to
the face are the terminal ones of the ophthalmic—namely,

the supra-orbital and supra-trochlear, which escape by the supra-orbital and supra-trochlear notches. Large inosculating vessels escape through several foramina in the bones of the face, from deep branches of the external carotid—viz., the infra-orbital, passing out of the infra-orbital foramen; from the mental foramen, and a large branch from the same source is found on the substance of the buccinator. The free inosculation of the arteries of the face renders ligature of both ends of a divided facial artery necessary, as the return circulation is very quickly re-established; and, in wounds, whether the results of accident or surgical interference, very accurate approximation of the edges must be obtained, for union takes place very rapidly, and distortion is not so easily remedied. The bloodvessels of the face are frequently subject to a nævoid condition.

The nerves of the face are derived from the three divisions of the 5th, and from the portio dura or facial part of the 7th cerebral nerve. The branches from the 1st division of the 5th are—the supra-orbital, supra-trochlear, infra-trochlear, lachrymal, and nasal. From the 2d—the infra-orbital, passing out of the infra-orbital foramen, and the subcutaneous malæ. The buccal from the same source emerges just in front of the anterior border of the masseter. From the 3d division—the masseteric, and the inferior dental from the foramen mentale. The facial forms a plexus in the parotid gland, after which it passes into a great many branches, and is supplied to the muscles of the face, having free inosculations with the branches of the 5th nerve. The infra-orbital nerve is peculiarly liable to neuralgic affections. Twitchings of the muscles of the face are connected with affections of the facial. Both nerves are occasionally palsied.

The *lymphatic ganglia* of the face are most thickly situated along the base of the jaw, on it and the bucci-nator muscle, others under the zygoma, and beneath, in, or upon the parotid gland. The ganglia around the mouth are sometimes affected with syphilitic induration after the application of the specific virus to the lips. The cellular tissue being very lax, and loosely attached to the subjacent structures, it is very liable to the infil-tration of fluids, or of air, as in wounds of the frontal sinuses or larynx. As the fasciæ of the face are very thin and ill-defined, abscesses in this region point early.

The congenital malformations consist of closure of its apertures, arrests of development, such as single and double hare-lip, frequently associated with cleft-palate. The aperture of the nostrils is occasionally single.

The facial relations of the *parotid gland* (anterior por-tion and socia-parotidis) are of importance so far as they are concerned in the removal of growths, and in opera-tions for salivary fistulæ. The gland is situated just in front of and below the ear, the deeper portion lying be-hind the angle of the jaw, limited above by the zygoma (*vide* Parotid Region). On the face it overlaps the mas-seter to a variable extent, having generally a small ac-cessory portion just in front of it, called the socia-paro-tidis. Its *duct* runs forward to the anterior edge of the masseter, and dips inward to open obliquely through the cheek, opposite the second molar tooth of the upper jaw. Its *course* is defined by a line extending from the upper border of the lobe of the ear to midway between the nostril and the angle of the mouth; and great care must be taken in operations on the face to avoid its division, as salivary fistula would be the result. The various structures which constitute the entire face will be found described in their surgical relations, with the deeper regions beneath them.

SURGICAL ANATOMY OF THE REGION OF THE NOSE.

The structure of the skin of the nose is very similar to that in other parts of the body, except that it is very thin and loosely connected with the subjacent parts. The hairs are but rudimentary, and the sebaceous glands very numerous and largely developed. These orifices show themselves as points more or less deep, most abundant on the alæ. Underneath is a layer of cellular tissue very adherent to the skin and subjacent musculo-fibrous tissues containing a little fat.

The *muscles* of the nose belong to those of expression, and are as follows: pyramidalis nasi, levator labii superióris et alæ nasi, dilator naris, compressor nasi, compressor narium minor, depressor alæ nasi.

The *arteries* of the nose are derived from the ophthalmic and the facial, the sides and dorsum being supplied by the nasal branch of the ophthalmic and the infra-orbital, the alæ and septum by the superior coronary, and by the lateralis nasi.

The *veins* terminate in the facial and ophthalmic.

The *nerves* are derived from the facial, infra-orbital, infra-trochlear, and a twig from the nasal branch of the ophthalmic of the fifth.

There are numerous *lymphatics*, which empty themselves into the submaxillary glands (lymphatic), and follow the course of the facial vein.

The mucous membrane lining the nostrils is continuous with the skin and that of the nasal fossæ.

The nasal cartilages, forming the softer portion of its framework, are five in number—viz., two upper lateral, and two lower lateral and the cartilage of the septum.

Each upper lateral cartilage is flattened and triangular in shape. Its anterior margin articulates with the cartilage of the septum. The posterior edge articulates with the nasal process of the superior maxillary and nasal bones. The lower edge is connected by fibrous tissue with the lower lateral cartilage.

The alar or lower lateral cartilages are two in number, and completely separate. They are of the form of a horseshoe, with the concavity posterior, and with the external limb longer than the internal. The convexity of these two cartilages is situated in the thickness of the lobe, on each side of the anterior inferior angle of the cartilage of the septum. The inner limb of this cartilage has its back against that of the opposite side and to the cartilage of the septum in the median line.

There are three or four small cartilaginous plates, situated in the tough membrane connecting the lower lateral (alar) cartilage with the nasal process of the superior maxilla—the *sesamoid*.

The alæ of each side are further composed of masses of cellular tissue placed below and behind the alar cartilages.

The bony framework is formed by the nasal bones, to which the external nose owes its form in a great measure; their method of articulation with the frontal and superior maxillary greatly determining its shape and dimensions.

These two oblong bones form by their junction along the middle line in front the " bridge " of the nose ; they articulate with the *frontal* and *ethmoid*, with the *superior maxilla* and with *each other*. The alar cartilages articulate by their lower edges.

The great vascularity of the nose and of the adjacent parts renders union after wounds very rapid; indeed, there are cases where the entire organ has been cut off, and been for some little while removed from the body, reuniting entirely after careful adjustment. Plastic operations for the restoration of the nose depend greatly for their success on the surgeon's ingenuity but also on so fashioning the flaps that they retain the vessels in their continuity, thus providing for their thorough nourishment.

SURGICAL ANATOMY OF THE NASAL FOSSÆ AND SINUSES OF THE NOSE.

Before exposing the contents of the nasal fossæ, an opportunity should be taken of examining the nares by means of the speculum, of performing the operation of passing probes into the nasal ducts and Eustachian tubes, of plugging the posterior nares, of introducing the fingers and instruments for the detection or removal of nasal and pharyngeal polypi, &c.

Dissection.—The saw is to be entered on one or other side of the crista galli, and to be carried gently downwards through the frontal and nasal bones, the cribriform plate, and a portion of the body of the sphenoid; the hard palate on the same side is next to be sawn through, the soft parts cut through with a scalpel in the same line, and the remaining portion of the body of the sphenoid divided. The nasal cavity being thus divided, one-half will show the septum and the other the meatuses. Each nasal fossa presents for examination a *roof*, *floor, internal wall, external wall,* and the *anterior* and *posterior nares.* The vertical diameter of the nasal fossæ is greater at the middle of the cavities than at the anterior or posterior parts, and the transverse diameter greater below than above.

The roof is formed by the nasal bones, the nasal spine of the frontal, the cribriform plate of the ethmoid, and the body of the sphenoid. It is to be noted that the entire roof is *not* horizontal, the cribriform plate only being so, and that it slopes downwards at front and back.

The floor is formed by the palate plates of the superior maxillary and palate bones.

The internal wall or septum is formed chiefly by the perpendicular plate of the ethmoid and by the vomer; the septal plane is further assisted by the nasal spine of the frontal, the crests of the nasal, superior maxilla, and palate bones. (The septum is rendered complete by the triangular cartilage, which projects forwards, assisting in giving shape and prominence to the nose.)

The outer wall is divided into the three meatuses by the projection from it of the three turbinated bones. It is formed by the nasal, the superior maxillary, the lateral mass of the ethmoid, and the lachrymal bones; posteriorly by the ascending plate of the palate, and the internal pterygoid plate of the sphenoid; the wall is completed by the lateral cartilages.

Meatuses.—The outer wall of each fossa is subdivided into three (sometimes four) irregular channels, termed meatuses—viz., superior, middle, and inferior. The bones entering into the formation of these meatuses are all those of the face, excepting the malar and inferior maxilla.

3

They are divided by the three turbinated bones.

The superior meatus lies beneath the superior spongy bone, and is the smallest, and has opening into it the posterior, ethmoidal, and sphenoidal cells. At the back is the spheno-palatine foramen, communicating with the spheno-maxillary fossa.

Fig. 2.

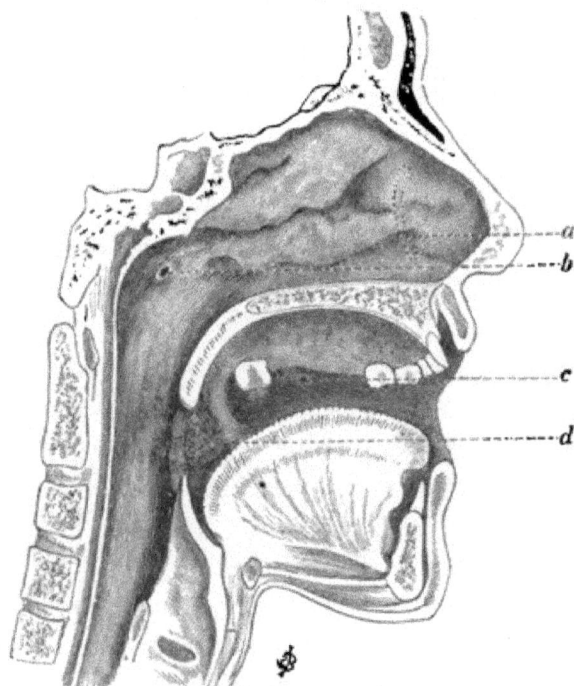

a. Position of nasal duct. b. Orifice of Eustachian tube. c. Orifice of Steno's duct. d. Tonsil between pillars of fauces.

The middle meatus lies beneath the middle spongy bone, and has opening into it in front, the frontal sinus (infundibulum), the anterior ethmoidal cells, and the opening of the antrum, which is almost impossible to find on the living body. (*Vide* Superior Maxillary Region.)

The inferior meatus, the largest and most important, lies beneath the inferior spongy bone; it extends almost the whole length of the fossa; its lower border, thick and rounded, descends almost to the floor, sometimes converting the meatus into a canal; anteriorly, where it joins the nasal process of the superior maxilla is the orifice of the nasal duct.

The opening into it is the *nasal duct.* This canal extends from the orifice in the meatus to the lachrymal sac (*vide* Orbital Region), and this orifice is situated about half an inch behind the ascending plate of the superior maxilla, and nearly opposite the centre of the under surface of the lower turbinated bone. It is somewhat valvular, owing to the folds of mucous membrane which pass into the aperture and which are continued up the tube. It takes a direction from *below, outwards, forwards, and upwards.*

To pass a probe into the nasal duct, the instrument should be first bent into the shape of an italic *f,* which should be passed first along the floor of the fossa, with its concavity towards the entrance; the point is next to be pushed gently beyond the ascending plate of the superior maxilla, and kept in close contact with the outer wall; then it is to be slightly rotated between the fingers, until the point presents upwards and outwards towards the eye; if the handle be now depressed, it should enter the canal.

The orifice of the *Eustachian* tube is a valvular aperture, situated on the inner surface of the internal pterygoid plate, and just above the posterior extremity of the inferior turbinated bone. *To introduce a sound or probe* into it the extremity of the instrument should be bent at an angle of about 60°, and passed along the floor of the nostril with the concavity downwards, then pushed backwards by the side of the septum until the mucous membrane of the back of the pharynx is reached; next it is to be slightly withdrawn, and rotated between the fingers, so as to bring its point upwards and outwards, which may be known to be in the orifice of the tube when it cannot be made to rotate easily.

The mucous membrane lining the nares completely covers the surfaces of the above-mentioned bony parts, terminating in front at its juncture with the skin. In color it is red, and its superior surface is studded with orifices of glands which secrete mucus. It is of variable thickness, and is very thin where it is prolonged into the several sinuses. It is thickest on the septum, especially so at its anterior half. It is moderately thick on the roof of the fossa, where it invests the proper bones of the nose and the cribriform plate of the ethmoid; it enters the sphenoidal sinus, and becomes very thin. It covers the anterior ethmoidal cells and the superior spongy bones; it sinks into the groove separating this bone from the sphenoidal sinus, and closes the sphenopalatine canal; and so it may be traced over each portion of the bony and cartilaginous structure to be continuous behind with the pharyngeal membrane and with the skin of the face in front.

Arteries.—This membrane is supplied by the sphenopalatine branch of the internal maxillary, which divides into two branches; the more internal being distributed to the septum, divides and passes towards the anterior palatine foramen; the external is distributed to the external parietes, and subdividing supplies the meatuses and spongy bones.

There are also branches from the superior dental (internal maxillary) to the antrum, from the infra-orbital, from the pterygo-palatine, and from the ethmoidal (ophthalmic) and facial.

Veins.—The veins form a peculiarly complicated plexus (rete nasi), which collect and pass forwards, terminating in the facial vein; others pass into the frontal vein, and another set pass through the spheno-palatine foramen into the plexus in the zygomatic fossa.

Nerves.—There are two sets of nerves distributed to the mucous membrane of the nose, viz. (α) those of special sense; (β) those of common sensation.

(α) The olfactory, derived from the olfactory lobe, penetrates in three layers the cribriform plate of the ethmoid bone, and is distributed to the ethmoidal and sphenoidal spongy bones, and to the upper part of the septum. (β) Those derived from the fifth are the nasal from the ophthalmic division, which passes to the anterior part of the mucous membrane, and leaves the cavity of the nose by passing between the lateral cartilage and the nasal bone (entering by the nasal slit). From Meckel's ganglion are given off the spheno-palatine branch for the septum and the external parietes; the posterior and inferior nasal from the anterior palatine, which is distributed to the posterior inferior portion of the external wall.

There is a point of considerable practical importance with regard to the relation borne by the velum pendulum palati to the posterior nasal apertures. If the mouth be opened, there is an involuntary disposition to breathe through it, and thus the palate applies itself closely to the walls of the pharynx, cutting off the communication between the nose and mouth. In syring-

ing out the nostrils, it will be found that if the mouth
be open and the nozzle of the instrument be introduced
into one of the anterior openings, the current of fluid
will wash out the entire nasal cavity, and pass out
through the other again. In the case of the introduction of the mirror in posterior rhinoscopy, the palate
must be forced forward by the emission of nasal sounds,
or drawn forwards by hooks or forceps constructed for
the purpose.

Nasal polypi of the fibro-cellular variety developed
in the submucous tissue of this region are covered with
ciliated epithelium, and are usually attached to one of
the superior turbinated bones. The more formidable
forms of growths generally project into the fossæ from
the antrum or base of skull. The posterior nares, oval
in form and of the same shape as their bony boundaries,
enter into the formation of a region termed the *nasopharyngeal,* situated midway between the nasal fossæ
and the pharynx. It is formed also by the body of the
sphenoid, the basilar process, and the pterygoid plates.

The basilar process is the usual locale of naso-pha-
ryngeal polypi, and its situation is readily recognized
by passing the finger behind the velum pendulum palati,
which frequently has to be divided in order to remove
such growths.

SURGICAL ANATOMY OF THE REGION OF THE ORBIT.

In the following short description of the orbital region
those parts only of the eyeball and appendages of the
eye which are the seat of the ordinary surgical operations are treated of. An account of those parts which
are contained within the globe, and of the operative

proceedings connected with them, are better referred to works devoted either to descriptive anatomy or to ophthalmic surgery, as the limits of the present volume scarcely admit of so special a subject.

External Orbital Region.

The external orbital region comprises that of the eyelids and the lachrymal apparatus.

The eyelids consist of the following layers: The most external is the skin, which is remarkably thin; next in order is the subcutaneous cellular tissue, very loose, and destitute of fat, continuous with that of the forehead, and very liable to infiltration, if the effusion of blood takes place beneath the tendon of the occipito-frontalis; then the orbicularis palpebrarum muscle, separated by the palpebral vessels from the tarsal ligament and cartilages. The arteries consist of the palpebral branches of the ophthalmic, which anastomose with the facial, temporal, infra-orbital, supra-orbital, and lachrymal. The veins, as a rule, accompany the arteries. The nerves are derived from the facial, the superior maxillary of the fifth and the third, which supplies also the levator palpebræ. The next layer is formed of cellular tissue and fat, particularly developed at the margin of the lids, where it is continuous with that of the orbit. The most internal layer is the palpebral conjunctiva, which covers in the Meibomian follicles. The tarsal cartilage of the upper lid is strengthened by the insertion into it of the levator palpebræ. The free border of the lids presents the orifices of the Meibomian and ciliary follicles, and their outer margin the cilia. The Meibomian glands are about thirty or forty in number, having corresponding ducts, on either side of which open a number of small cul-de-sacs. The office of their secretion is to prevent adhesion of the eyelids; the ciliary follicles are subject to dilatation and suppuration, forming *stye*. The inner angle presents the tendo-oculi, the inosculation of the angular and ophthalmic arteries and their veins, the lachrymal sac, puncta lachrymalia, and in the canthus the caruncula lachrymalis, having on its outer side the plica semilunaris.

The *conjunctiva*, after its reflection over the eyeball, becomes transparent, and its bloodvessels are invisible, unless conjunctivitis is present. These vessels are arranged as a network over the entire surface of the globe, and can be caused to slide over the subjacent cornea or sclerotic, owing to the cellular membrane existing between them. It will be observed in the inflammatory condition, that the direction of the bloodvessels of the sclerotic are arranged radially, and are pinkish in color, owing to their lying in its dense substance, whereas the conjunctival vessels are scarlet.

The lachrymal apparatus, which includes the lachrymal gland, the lachrymal ducts, the conjunctival surface of the eyeball, the canalicula, the lachrymal sac and its duct (nasal duct), is situated partially within the orbital cavity, partially in the eyelids, and partially in the nose.

The lachrymal gland lies in a depression situated at the external and superior aspect of the orbital fossa. It is inclosed in a fibrous capsule, and generally consists of two portions—an orbital, the larger, and an anterior or palpebral; the outer margin of the aponeurosis of the levator palpebræ muscle forming a partial separation of these two portions. *Its relations* are as follows: *Superiorly,* the periosteum of the orbit (dura mater), *inferiorly,* the eyeball, and superior and external recti muscles; *anteriorly* it is closely adherent to the posterior aspect of the upper lid. *Within* it are the lachrymal branch of the ophthalmic artery, inosculating with the orbital branch of the internal maxillary, with their accompanying veins; the lachrymal branch of the ophthalmic division of the fifth nerve, which inosculates with the orbital branch of the superior maxillary division of the fifth. These structures enter at its posterior and external margin. The ducts of the lachrymal gland, which are about ten in number, are so arranged that they open in a row on to the conjunctiva, where it is reflected from the upper lid to the globe.

The lachrymal canals commence at the puncta lachrymalia, which are seen, on everting the lid, as the centres of small eminences, situated about a quarter of an inch from the inner angle, on the inner aspect of the margin of the lid. These minute openings are kept in contact with the conjunctival surface of the globe by the action of the tensor tarsi, so that they always lie in the current of the tears. In each canaliculus, immediately below the punctum, is a small cul-de-sac, beyond which, after inclosing the caruncula, the canals join previous to entering the lachrymal sac. Their posterior portion is subconjunctival, a circumstance of considerable practical value in the operation of reinstating the course of the tears when from any cause the orifices of the puncta do not perform their function of receiving the tears. That portion which is common to both canaliculi is bound down by the tendo-oculi. The process by which the continuous flow of tears between the lachrymal gland and the nose is kept up, is not entirely clear; it is very

4

probably in a great measure owing to suction caused by exhaustion in the nasal duct.

Fig. 3.

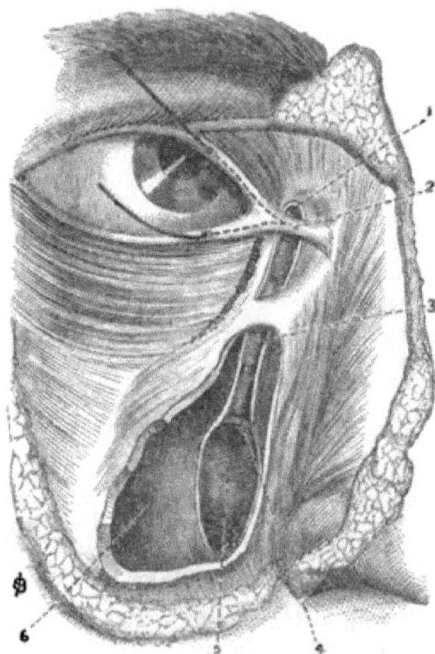

Lachrymal apparatus and nasal duct. 1. Lachrymal sac. 2. **Tendo-?culi.**
3. Valvular folds in nasal duct. 4. Orifice of nasal duct. 5. Lower turbinated
bone. 6. Inner wall of antrum. (Bristles introduced into the puncta lachrymalia.)

The lachrymal sac is the superior dilated extremity of the nasal duct, and consists of a tube of fibro-elastic tissue, lined with epithelium, continuous through the puncta with the conjunctiva, and by means of the nasal duct, with the mucous membrane of the nose. It is situated at the internal angle of the eye, and is lodged in a hollow formed by the os unguis and the nasal process of the superior maxillary. It is *covered* in by the skin, subcutaneous cellular tissue, orbicularis muscle, by the tendo-oculi, and the internal portion of the palpebral aponeurosis. The exact situation of the sac can be felt with the finger, and the best landmark is the anterior lip of the lachrymal groove, surmounted by a small tubercle of bone, formed by the external border of the nasal process of the superior maxilla, or if both the lids be abducted from the mesial line, tension is made on the tendo-oculi, which will show itself as a flat cord immediately over the sac, bisecting it.

To introduce a Probe into the Nasal Duct by the punctum lachrymale.—It is frequently necessary to pass a probe into the nasal canal through the punctum; the lower lid is to be everted, when the punctum will be seen about two lines from the inner angle on a small papilla. The probe is first to be introduced vertically, and pushed downwards for a short distance, when the hand is to be depressed, and the probe pushed inwards until arrested by the os unguis, then raised again vertically; when the slightest pressure will cause it to traverse the lachrymal sac and enter the nasal canal, pushing it downwards, backwards, and inwards.

When suppuration takes place in the lachrymal sac, and ulceration through the integument follows, *lachrymal fistula* is the result.

The tears may be prevented passing into the puncta, owing to their being obstructed, or to eversion of the lid from some cause, such as ectropion, as a result of a burn or other injury, or cicatrization after syphilitic ulceration.

Internal Orbital Region.

Dissection.—To expose the contents of the orbit (the skull cap having been removed) a saw should be entered

through the frontal bone, first in a line with the inner
angle and the optic foramen, and again in a line with
the outer angle and sphenoidal fissure; a few taps on
the orbital plate with a hammer will break it through,
and the triangle of bone can be readily tilted forwards
by a blow on the margin of the skull; the ring round
the optic foramen should be retained, as the muscles are
attached to it.

Boundaries.—The bony walls of the orbit are formed
as follows:

The roof, concave, directed downwards and forwards,
by the orbital plate of the frontal bone in front, and by
the lesser wing of the sphenoid behind. *The floor,* nearly
flat, by the malar, superior maxillary, and orbital plate
of palate. *Outer wall,* concave, by the greater wing of
sphenoid and malar bone. *Inner wall,* flat, by the lachrymal, os planum of ethmoid, and sphenoid. Regarding
the shape of the orbit as nearly conical, its base is nearly
quadrilateral, and at its several angles are found the sutures of its component bones. At its external, that of
the external orbital process of the frontal with the malar;
at its internal, that of the frontal with the lachrymal
and nasal process of superior maxilla; below, that of the
malar with the superior maxilla.

Its apex corresponds to the optic foramen and sphenoidal fissure.

The spheno-maxillary fissure is found on the floor of
the orbit. The continuity of this fissure with the
spheno-maxillary fossa explains the protrusion of the
eyeball from processes of tumors passing through it
from the spheno-maxillary region.

Contents.—The periosteum of the orbit is formed by
the dura mater, and enters the cavity by the optic fora-

men and anterior lacerated fissure. Each surface splits into two layers—one continuous with the pericranium at the upper margin of the circumference, or with the periosteum of the face, and the other which forms the palpebral ligament of the lids. *The muscles* which act upon the globe are six in number—namely, the four recti and two obliqui, and one muscle acting upon the upper lid, the levator palpebræ. These muscles are inclosed in fibrous sheaths derived from the orbital aponeurosis. At their insertion their tendons become expanded upon and continuous with the sclerotic. The complete division of the aponeurotic sheaths as well as of the tendons is necessary in the operation for strabismus, as these investments if left undivided still exert considerable power over the globe, owing to their completely inclosing both muscle and tendon. The recti tendons are inserted into the sclerotic about a quarter of an inch behind the cornea. The upper and lower eyelids are united to the sheaths of the superior and inferior recti muscles by an offset of the palpebral aponeurosis, forming the posterior boundaries of the superior and inferior oculo-palpebral cul-de-sacs. A knowledge of the arrangement of these aponeurotic expansions and of their situations is of some considerable importance in the diagnosis of intra-orbital injury or disease, as by their attachment and situation they facilitate or impede the course taken by blood or pus. Ecchymosis beneath the conjunctiva is almost invariably a symptom of fracture of the roof of the orbit after injury of the head.

The eyeball and its vessels and nerves lie in a mass of fat and cellular tissue, which serves as a cushion for the optic nerve and for the globe in its various movements, and as a support for its accessory structures. This cellulo-fatty mass is continuous with the cranial cellular tissue, and with that of the zygomatic and spheno-maxillary fossæ.

FIG. 4.

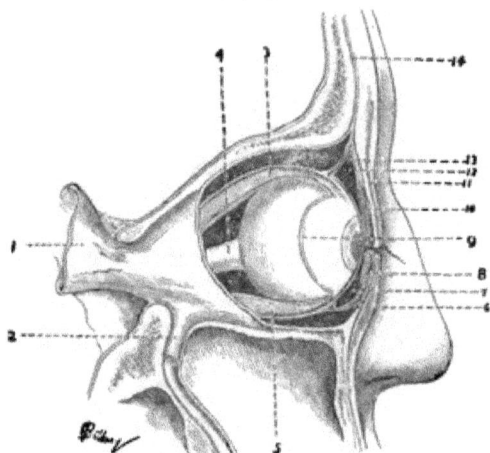

Aponeuroses of orbit. (Altered from BÉRAUD.) 1. Dura mater. 2. Prolongation of dura mater into the posterior palatine canal. 3. Superior rectus inclosed in its sheath. 4. Optic nerve. 5. Inferior rectus in its sheath. 6. Process of aponeurosis of inferior oblique attached to the palpebral aponeurosis. 7. Inferior oculo-palpebral cul-de-sac. 8. Inferior tarsal cartilage. 9. Ocular portion of orbito-ocular aponeurosis. 10. Superior tarsal cartilage. 11. Palpebral aponeurosis. 12. Superior oculo-palpebral cul-de-sac. 13. Process of aponeurosis of superior oblique attached to the palpebral aponeurosis. 14. Periosteum of frontal bone continuous with that of the orbit.

The arteries are derived from the ophthalmic branch of the internal carotid, which enters the optic foramen below the nerve, and forms free anastomoses with the temporal, facial, and internal maxillary.

The veins generally accompany the arteries, and terminate in a large vein, the ophthalmic, unprovided with valves, which, after being formed both without and within the orbit, passes as a trunk between the two heads of the external rectus muscle, and enlarges into the cavernous sinus.

The nerves are—the optic, the third, the fourth, the ophthalmic division of the fifth and the sixth, with their branches, and from the lenticular ganglion are given off the ciliary (short). Paralysis of the third nerve or motor-oculi, causes *ptosis*, or a dropping of the upper lid, external strabismus, dilatation, and immobility of the pupil. Paralysis of the fourth nerve, or patheticus, causes impossibility of rotation of the eyeball, and diplopia; in paralysis of the sixth, or abducens, the eyeball is turned inwards.

Relations of Parts within the Orbit.—A good idea of the actual relations of the contents of the orbital cavity looked upon as a cone may be obtained by regarding the eyeball and optic nerve as occupying very nearly its axis, and the muscles, vessels, and nerves as placed superiorly, inferiorly, externally, and internally to them. *A needle passing through the axis of the eyeball* from its anterior surface backwards would traverse successively—(1) the ocular conjunctiva; (2) the four layers of the cornea—viz., anterior elastic lamina, the cornea proper, the posterior elastic lamina, the posterior epithelium; (3) the anterior chamber; (4) the pupil; (5) the anterior layer of the capsule of the lens; (6) the lens; (7) the posterior layer of the capsule of the lens; (8) the anterior portion of the hyaline membrane; (9) the vitreous humor; (10) the posterior portion of the hyaline membrane; (11) the three layers of the retina—viz. (α), Jacob's membrane (rods and cones); (β) the granular layer; (γ) the fibrous layer; (12) the choroid; (13) the sclerotic.

The relations of the globe and optic nerve, considered as occupying the axis of the orbital cavity, successively exposed by dissection, from either surface inwards, would be as follows:

Superiorly, from above downwards (omitting the dura mater): (1) The frontal vessels and nerves, and behind in the same plane the fourth nerve; (2) the levator palpebræ muscle; (3) the rectus superior; (4) the superior set of the muscular branches of the ophthalmic artery, and the superior division of the third nerve; (5) the nasal nerve and ophthalmic artery, and the ciliary vessels and nerves.

Inferiorly, from below upwards: (1) The inferior rectus, and the inferior oblique muscles; (2) the inferior division of the third nerve, and its branch to the lenticular ganglion; (3) the inferior set of muscular vessels.

Internally, from within outwards: (1) The superior oblique and internal rectus muscles. (2) The termination of the fourth nerve, the nasal nerve, the ophthalmic artery and vein, and the anterior ethmoidal artery.

Externally, from without inwards: (1) The lachrymal vessels and nerve, and the lachrymal gland. (2) The external rectus muscle, between the two heads of which pass, both divisions of the third nerve, the nasal of the fifth, the sixth nerve, and the ophthalmic vein. (3) The nasal nerve, lenticular ganglion, and ophthalmic artery.

The structures divided in the operation of extirpation of the globe are—the conjunctiva, the subconjunctival tissue, the tendons of the recti and obliqui with their aponeurotic sheaths, the optic nerve, and the long and short ciliary vessels and nerves.

The structures divided in the operation for strabismus are—the conjunctiva, subconjunctival tissue, and rectus tendon with its aponeurotic sheath.

SURGICAL ANATOMY OF THE SUPERIOR MAXIL-LARY REGION.

This region may be regarded as that occupied by the superior maxilla and the tissues covering it. The superior maxilla is a bone of great surgical interest, on account of the many diseases to which it is liable; hence its position, relations, and connections are of the highest practical importance.

The structures exposed in their order on dissecting down upon the superior maxilla are—the skin and superficial fascia; the lower fibres of the orbicularis palpebrarum; the facial and infra-orbital vessels and nerves; the zygo-matici and the levator labii superioris; Steno's duct; the transverse facial artery; the buccal vessels and nerves; lymphatics, and the buccinator muscle. In the hollow between the anterior border of the masseter and the buccinator muscle is a large quantity of fat and cellular tissue, which contributes, either by its excess or deficiency, to the general contour of the face.

Articulations of the Superior Maxilla.—Articulating with its fellow, it forms the whole of the upper jaw; besides this, it articulates with the frontal, ethmoid, nasal, inferior turbinated, palate, vomer, malar, and lachrymal. The sutures it forms with those bones which enter into the formation of the face are very strong and difficult to separate, so much so, that in excision it will generally be found more satisfactory to divide it or its associate near to the articulation than to attempt to wrench them apart at the sutures. The processes requiring division in its extirpation are the palatine, nasal, and malar. Each bone assists in the formation of three cavities,—the mouth, nose, and orbit; of two fossæ, the zygomatic and

spheno-maxillary; and of two fissures, the spheno-maxillary and pterygo-maxillary. The relation of these
fissures and cavities to the body of the bone is of great
importance in the process of its removal. The apex of
the antrum corresponds on the face to its malar process;
the base of which looks inwards to the outer wall of the
nose; and its roof is formed by the orbital plate, and its

FIG. 5.

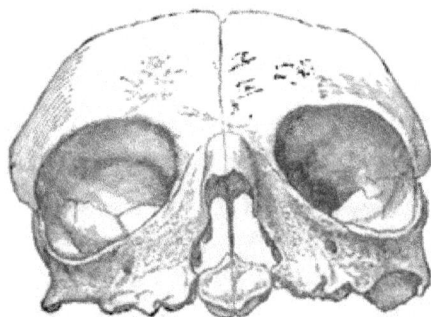

FIG. 6.   FIG. 7.

Fig. 5 shows the defective development of the superior maxillary bones, from
a fœtus seven months old; and Figs 6, 7 show the central mass formed of two
portions. In these cases, particularly in Fig. 5, the fact of the central portion
being composed of two intermaxillary bones is well seen, the fissure being in the
mesial line exactly under the nostril. (FERGUSSON.)

floor by the alveolar process. The walls of the antrum
are very thin, so that growths or collections of fluid
readily cause a bulging of its parietes and protrusions
into neighboring cavities or fissures. The fangs of the
first and second molar teeth project into its floor, hence
the importance of extracting one of these teeth and per-

forating its socket before interfering with any doubtful tumor connected with the cavity.

Besides the growths which are developed in the antrum, the bone is surgically interesting as being subject to an arrest in its development, known as *fissured or cleft palate*, frequently associated with a similar one in its appendage, the upper lip, termed *hare-lip*. During its development that portion which carries the incisor teeth is a separate segment, and if this segment be ununited the result is a deep fissure, extending backwards into the palate; occasionally these segments in both bones are thus disconnected, in which case they both hang from the end of the vomer, leaving a chasm in the roof of the mouth, a condition usually associated with a double hare-lip.

Structures divided in Excision of the Upper Jawbone.— Supposing the incision through the upper lip, along the ala of the nose, towards the inner angle of the orbit and along its lower margin to be adopted; first are the tissues composing the upper lip—viz., the integument, the orbicularis oris muscle, the cellular tissue containing the labial glands, the coronary vessels, facial, and branches of the second division of fifth nerves, and the mucous membrane. Next carrying the incision along the ala and side of the nose, the integument, fascia, levator labii superioris alæque nasi, with its aponeurosis, the angular vessels and branches of the infra-orbital and facial nerves. The incision along the lower border of the orbit divides the integument, aponeurosis, orbicularis palpebrarum muscle, the vessels of the lower eyelid, and the orbital fascia; and if a portion of the floor of the orbit be removed, the tendon of the inferior oblique, and by turning back the flap inclosed by these incisions, the attachments to the bone of the following muscles,—orbicularis, levator labii superioris et alæ nasi, levator labii superioris, compressor naris, depressor labii superioris, levator anguli oris, buccinator, infra-orbital vessels and nerve, and facial vessels and nerves. The advantage of this method of external incision is that the vessels are divided near their termination, and not through their larger branches, and the duct of the parotid is left entire, without the risk of salivary fistula, besides leaving an almost unnoticeable cicatrix, by following the natural furrows of the face.

In the second stage of the operation, an incisor tooth being extracted, the gum, alveolar process, and structures forming the hard palate—malar and nasal processes, with a portion of the floor of the orbit. In the subsequent dislocation of the bone, the internal maxillary artery, with its vein and the branches after they have gained the pterygo-maxillary fossa, and the posterior palatine nerves are divided.

SURGICAL ANATOMY OF THE REGION OF THE SOFT PALATE AND TONSIL.

The soft palate, which is suspended obliquely from before backwards from the posterior border of the palatine arch, or hard palate, is a curtain consisting of mucous membrane, muscular and fibrous tissue, vessels and nerves, forming an incomplete septum between the nasal and buccal cavities, serving to prevent the food from passing upwards into the nasal fossæ, helping to push it downwards into the pharynx during deglutition, and also acting upon the quality of the voice. Its movements are elevation, depression, and transverse tension. It is concave anteriorly, and its inferior anterior border presents two semilunar margins, the edges of which pass downwards to the sides of the tongue, united in a central raphé, from which depends a prolongation, the uvula.

In structure the soft palate consists anteriorly and inferiorly of a mucous membrane, thickly studded with muciparous glands, continuous with that of the posterior region of the mouth, and posteriorly and superiorly of a second membrane, continuous with that of the nasopharyngeal region. Between these mucous layers is a musculo-tendinous one, consisting of portions of the following pairs of *intrinsic* muscles,—the levatores palati and the tensores or circumflexi palati; and of *extrinsic* muscles, the palato-glossi and the palato-pharyngei. Some delicate muscular fibres are to be found in the uvula. The order in which the structures enter into the formation of the velum is as follows: From before backwards, (1) the anterior layer of mucous membrane; (2) the aponeurosis of the tensores palati, with which is blended the attachments of the palato-glossi and palato-pharyngei; (3) the levatores palati, uniting in the median raphé; (4) the posterior layer of mucous membrane.

The pillars of the fauces are formed by the divergence of the palato-glossus and palato-pharyngeus, and include a triangular interval, the base being downwards, in which is situated the tonsil. The posterior pillars formed by the palato-pharyngei are nearer each other than the anterior, formed by the palato-glossi.

The space between the palatine arches of both sides is called the *isthmus of the fauces*, and is bounded above by the free margin of the palate, below by the dorsum of the tongue, and laterally by the pillars of the fauces and tonsils.

A correct knowledge of the attachments and actions

of the muscles of the soft palate is of great importance, with a view to the successful performance of operations for the relief of fissures or clefts in it.

The fibres of the levator palati pass downward and inward, spreading out on the velum as a layer, which is embraced by the two planes of fibres of the palato-pharyngeus and unites with its fellow of the opposite side. The tensor palati ends in a tendon, which is reflected horizontally round the hamular process of the sphenoid, and after spreading out is inserted into the aponeurosis of the velum, below the levator palati, and

Fig. 8.

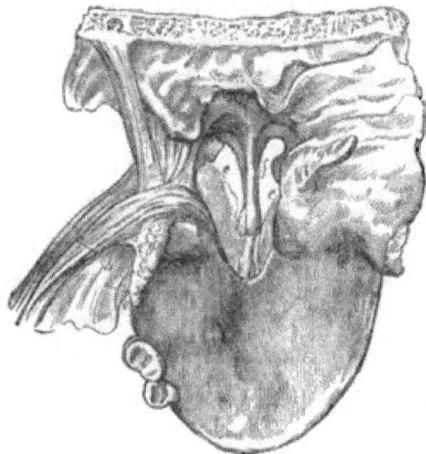

This figure represents the posterior nares and upper surface of the soft palate. *a.* Levator palati; the dark line shows where it should be cut across. *b.* The inner bundle of fibres of the palato-pharyngeus; the dark line indicates its place of division. *c.* The palato-glossus, with the mark for incision, when necessary. *d.* The tensor palati, in relation with the cartilaginous extremity of the Eustachian tube. *e.* The posterior extremity of the inferior turbinated bone. *f.* The septum. *g, g.* The uvula on each side stretched apart. (FERGUSSON.)

into the palate bone. *The position of the hamular process,* an important guide in the performance of operations, can be felt distinctly in the substance of the soft palate, internal to and slightly posterior to the last molar tooth.

The action of the palatine muscles upon a fissure existing in the velum would obviously produce a separation of its margins, and it has been shown that the muscular action by which these margins are brought together is caused by the upper semicircular border of the superior constrictor of the pharynx, and that the muscles to be divided in the operation of *staphyloraphy* or stitching up the fissure, are the levatores palati and the palato-pharyngei, the upper expanded fasciculi of which are divided into two parts by the levatores palati, and if necessary the palato-glossi. (Fergusson.)

The levator palati is to be divided on both sides by putting the undeveloped velum upon the stretch, when a double-edged knife is passed through the soft palate, just on the inner side of the hamular process, and above the line of the levator palati.

Another method of dividing the levator palati is by passing a knife curved on the flat through the fissure and behind the flap, its edge making an incision half an inch long, half way between the hamular process and the orifice of the Eustachian tube, and perpendicular to a line drawn between them.

The palato-pharyngeus is to be divided by cutting through the posterior pillars just below the tonsil. Occasionally the palato-glossus requires division.

The tonsils or *amygdalæ* are two small glandular bodies, varying in size in different individuals, situated between the anterior and posterior pillars of the fauces; they are in relation, *externally* with the superior constrictor, and by it separated from the internal carotid and ascending pharyngeal vessels; *below* they rest on the side of the base of the tongue. The position of the tonsil corresponds with the angle of the inferior maxilla, at a point

nearly opposite the root of the alveolar process of the
second molar tooth.   Under certain circumstances the
carotid artery is in danger of being wounded, such as in
excision of the gland or the evacuation of pus, when by ·
its enlargement it is brought still more closely into connection with the vessels, but if the precaution be taken
of lifting it well from its bed forwards and inwards
before the knife is applied for its removal, the risk in

FIG. 9.

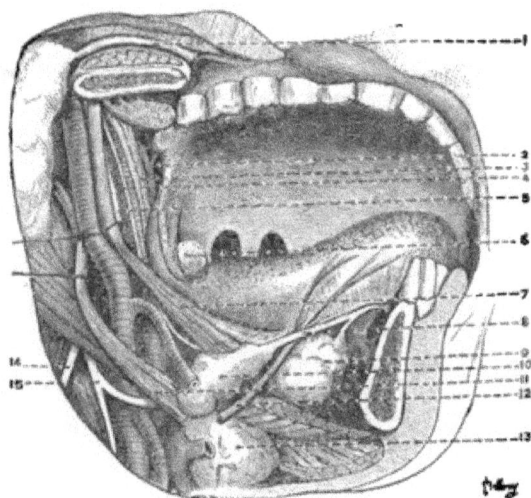

Relation of the right tonsil viewed laterally, the half of the lower jaw having
been removed. 1. Steno's duct crossing the masseter and opening into the buccal
cavity. 2. Ascending pharyngeal artery. 3. Stylo-pharyngeus muscle. 4. Pharyngeal branch of vagus. 5. Glosso-pharyngeal nerve. (Behind which is seen the
internal carotid artery.) 6. Tonsil lying between the pillars of the fauces. 7.
Stylo-glossus muscle, hooked aside. 8. Wharton's duct. 9. Sublingual gland.
10. Superior lobe of submaxillary gland. 11. Stylo-hyoid muscle, hooked aside.
12. Gustatory nerve. 13. Submaxillary gland. 14. Spinal accessory nerve. 15.
Hypoglossal nerve.

this instance is avoided ; and in the latter, care must be
taken not to push the knife forwards in the line of the
angle of the jaw, but backwards into the tumor towards

the spine, and so allow it to cut its way out, towards the median line of the body.

The *vessels* of this region are derived from the ascending pharyngeal of the external carotid, the ascending palatine and tonsillitic of the facial, the dorsalis linguæ

Fig. 10.

Sketch of the relations of the left tonsil viewed from above. 1. Superior cervical ganglion. 2. Internal carotid artery and jugular vein. 3. Digastric muscle. 4. External carotid artery. 5. Glosso-pharyngeal nerve (drawn too thick). 6. Stylo-pharyngeus muscle. 7. Stylo-glossus muscle. 8. The tonsil. 9. Section of the pharynx and its mucous membrane.

of the lingual, and the descending palatine of the internal maxillary ; and the *nerves*, from the glosso-pharyngeal and Meckel's ganglion.

SURGICAL ANATOMY OF THE PAROTID REGION.

The boundaries of the parotid region are somewhat difficult of definition, partly on account of the irregularity of the gland, and partly on account of its belonging both to the cranium and to the neck. The following appear to be the simplest: *In front*, the posterior

54 SURGICAL ANATOMY OF

border of. the **ramus** of the jaw; *behind*, the mastoid process of the temporal bone, the cartilage of the ear and the edges of the sterno-mastoid and digastric muscles; *above*, the zygomatic arch; and *below*, an imaginary line drawn horizontally backwards and inwards from the angle of the jaw to the styloid **process, by** the stylo-hyoid and stylo-maxillary ligaments, and the process of cervical fascia passing from the sterno-mastoid to the jaw. The dimensions of this region obviously vary with the several movements of the lower jaw; moreover, there are certain differences in its size with respect to the age of the individual; thus, in the infant, the region is broader in proportion below, on account of the obliquity of the jaw and the non-development of its angle, and bulges externally, on account of the quantity of fat and lymphatics contained within it; again, in old age, in the edentulous state, the base of the region becomes broader, owing to the falling forward of the jaw.

Dissection.—On removing the integument covering the parotid gland, it will be seen to be enveloped in an incomplete capsule derived from the cervical fascia which separates it from neighboring structures.

The relations and connections of the parotid gland are, *externally and superficially* the lymphatics, the platysma myoides, some few branches of the superficial cervical plexus, and the integument; *anteriorly*, the posterior border of the ramus of the jaw, the external and internal pterygoid muscles, between which lies a process of the gland; *inferiorly and posteriorly*, the mastoid process, the sterno-mastoid, posterior belly of digastric, styloid muscles, transverse process of atlas, internal jugular vein, internal carotid artery, eighth pair of nerves, hypoglossal nerve, and the superior cervical ganglion of the sympathetic.

The substance of the gland contains so many important structures that operative proceedings connected with it are rendered excessively difficult and hazardous. The external carotid artery traverses its posterior part, giving off its anterior and posterior auricular and superficial temporal branches. Behind the external carotid is the external jugular vein, which receives numerous branches in its substance. A quantity of lymphatics

FIG. 11.

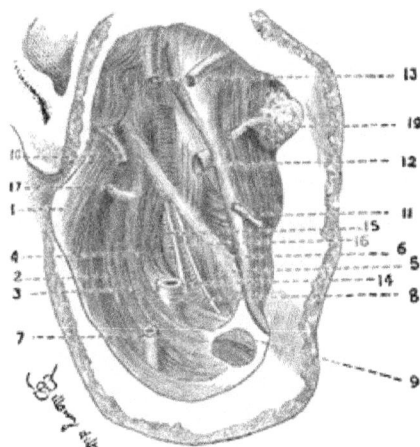

Sketch of the deep relations of the right parotid gland (the gland itself has been removed and the ramus of the lower jaw drawn forward). 1. Remains of aponeurosis of gland. 2. Digastric muscle. 3. Stylo-hyoid muscle. 4 Stylopharyngeus (drawn aside). 5. Stylo-glossus muscle. 6. Stylo-maxillary ligament. 7. External jugular vein. 8. External carotid. 9. Lingual artery seen through an opening in the aponeurosis. 10. Posterior auricular artery. 11. Transverse facial artery. 12. Internal maxillary artery. 13. Anterior and middle temporal arteries. 14. Internal carotid. 15. Hypoglossal nerve. 16. Glosso-pharyngeal nerve. 17. Trunk of facial nerve. 19. Steno's duct.

are also found in relation with the gland. The superficial lymphatic ganglia receive the vessels of the scalp; those within the gland, the vessels from the eyebrows, lids, and cheeks; and the deepest, which accompany the internal carotid, are the vessels of the temporal and maxillary regions. The facial nerve, after it has passed through the stylo-mastoid foramen, enters the posterior and inferior portion of the parotid, and thence spreads out into several large plexiform branches (*pes anserinus*), after which it ramifies amongst the muscles of the face. The auriculo-temporal branch of the inferior maxillary nerve also enters the gland after having passed behind the neck of the jaw, and forms inosculations with the facial. The auriculo-parotidean branch of the cervical plexus enters the gland anteriorly and inferiorly, inosculating with the preceding.

The existence of the tough fibrous investment which incloses the gland almost entirely and binds it so tightly in its place, accounts for the intense pain in inflammation, as in *parotitis*, or abscess. Wounds, or the results of abscess in the substance of the gland, may give rise to *salivary fistulæ*, which are frequently very troublesome to close; and in the removal of tumors connected with it, or in its neighborhood, there is, of course, great danger of severe hemorrhage, and of wounding the facial nerve, thus causing paralysis of the facial muscles. The surgical relations of the duct of the parotid have been already considered (*vide* Face). The external carotid may be compressed against the styloid process in the adult, but it is impossible in the child, owing to the undeveloped state of that portion of the bone.

SURGICAL ANATOMY OF THE PTERYGO-MAXIL-LARY REGION.

The surface markings of this region are the bony prominences of the zygoma and lower jaw, and the contour of

the masseter muscle, the tendinous intersections of which
are very evident during its action. If the finger be
passed into the mouth, the superior border of the lower
jaw, the internal aspect of its ramus, and its anterior
border can be felt; and if the jaws are apart, its coronoid process; the mutual relations of which should be
carefully noticed.

Dissection.—The superficial dissection of the face being
completed, and the facial nerve, the transverse facial
vessels, with the duct of the parotid, cut and turned
forwards, that portion of the parotid which lies upon the
masseter is to be turned back towards the ear, and the
masseter itself exposed. Two sets of fibres will be seen
—the anterior forming the greater bulk of the muscle,
and behind and below these some oblique fibres, which
are inserted beneath the anterior. The origin of these
fibres is on the under and inner surface of the zygomatic
arch. It is to be noticed that the zygoma is entirely
subcutaneous. Next, the extremities of this process are
to be divided with a saw, and the included portion of
bone with the origin of the muscle turned downwards
upon the ramus of the jaw, noticing the attachment of
the temporal fascia to its upper border, and taking care
not to divide the masseteric nerve and artery, which
enter the under and upper aspect of the masseter through
the sigmoid notch, or to injure the external lateral ligament of the temporo-maxillary articulation. The ramus
of the jaw is next to be divided transversely, about
three-quarters of an inch below the notch, taking care
not to injure the inferior dental vessels and nerve which
enter the bone on its inner surface. The neck of the
condyle should then be nipped through, and the portion
of bone (including the coronoid process and a part of

the ramus) carefully lifted up, with the attached temporal muscle; some fibres of this muscle must be divided in order to do this, and as the buccal nerve and
artery usually lie immediately beneath them, some caution is necessary in order to avoid their division.

Contents.—Immediately beneath the bone thus raised,
are seen a portion of the internal maxillary artery, the
external pterygoid muscle, and emerging from its lower
border the gustatory branch of the third division of the
fifth, the mylo-hyoidean branch of the inferior dental
nerve, part of the internal pterygoid muscle, the trunk of
the inferior dental and deep temporal nerves, the internal lateral ligament of the temporo-maxillary articulation, and in front of the internal pterygoid the posterior
portion of the buccinator.

The vessels of this region are, in the *superficial dissection,* the transverse facial or external maxillary, which
arises from the external carotid in the parotid gland,
just above the angle of the jaw, and in relation with the
portio dura and Steno's duct, crosses the masseter a little below the zygoma.

In the *deep dissection* the internal maxillary artery
commences at the outer border of the neck of the jawbone, lying in the first part of its course behind it, and
in front of the internal lateral ligament; it then curves
forwards to the lower border of the external pterygoid
muscle, lying on it, and, generally disappearing between
its two heads, passes into the pterygo-maxillary fossa.
The trunk of this vessel or its descending branch, the
inferior dental, are usually divided in resection of the
bone. Its tortuous course and variable position render
it the more liable to be wounded.

The nerves are derived from the inferior maxillary di-

vision of the fifth, which passes into the region through
the foramen ovale of the sphenoid.   The nerve consists
of two portions—a muscular, distributed to all the muscles of mastication — viz., masseteric, deep temporal,
pterygoid, buccal; and a sensory—the inferior dental,
auriculo-temporal, and gustatory.   The relations of the
salivary glands to the body of the inferior maxilla are
of considerable importance with regard to operations on
it.   The parotid envelops the posterior border of its
ramus, and passes behind its neck as far as the styloid
process of the temporal bone; the submaxillary gland is
partially lodged in a fossa, below the attachment of the
mylo-hyoid, which during flexion of the head is concealed beneath, and in extension is considerably disen-
gaged from the bone—facts to be considered in operations in the region of the upper part of the neck.

The temporo-maxillary articulation is formed between
the glenoid cavity of the temporal bone and the condyle
of the lower jaw.   Interposed between the bones is a
biconcave interarticular fibro-cartilage, above and below
which is a synovial membrane.   The external portion of
the circumference of this cartilage is connected with the
external lateral ligament, internally with the capsular
ligament, whilst a portion of the tendon of the external
pterygoid muscle is inserted into it in front control-
ling its movements   *The ligaments* are the external
lateral, attached to the tubercle on the zygoma and to
the external surface of its neck, which is covered over
by the parotid gland; and the internal lateral, attached
to the spine of the sphenoid and to the inner margin of
the dental foramen.   The external pterygoid muscle is in
relation with this ligament above, the internal maxillary
artery lies between it and the neck of the jawbone, and

between it and the ramus of the jaw are the inferior
dental vessels and nerve; and on its inner side is the internal pterygoid muscle.   The stylo-maxillary ligament,
really a slip of the deep cervical fascia, is attached to
the styloid process of the temporal bone and to the inferior angle of the lower jaw, it separates the parotid
from the submaxillary gland, and gives origin to some
fibres of the stylo-glossus muscle.

Owing to the numerous movements which the muscles of mastication are capable of causing, and to the
comparative laxity of the ligaments of the articulation,
the lower jawbone is liable to partial or complete *dislocation*.   The condyles glide forward, carrying the inter-
articular fibro-cartilages with them, upon the eminentiæ
articulares, in such conditions as yawning or laughing,
or masticating large morsels; the combined action of the
masseter and internal pterygoid muscles drags them
under the zygomatic arches, whilst the temporal muscles
drag the displaced bone upwards.   The obstacle to reduction appears to be, that in most cases the coronoid
process is, as it were, locked in front of the malar tubercle.   Partial luxation of one condyle is of common oc-
currence, and occasionally a portion of that process of
the parotid which wraps round the neck of the jaw is
included between the opposing surfaces, causing severe
pain and inconvenience.   The principle upon which reduction is effected is by introducing some solid body
between the molar teeth so as to form a fulcrum, whilst
the power is applied at the symphysis, at the same time
that the angles are depressed.

SURGICAL ANATOMY OF THE LINGUAL REGION.

The tongue completely occupies the cavity of the mouth when it is shut, and is attached for the posterior two-thirds of its volume to the hyoid bone and inferior maxillary bone, by its extrinsic muscles and membranes; by means of the stylo-glossus it is attached to the styloid process, and by the palato-glossus to the palatine arch. It is at this portion of the organ that the nerves and arteries enter and its veins leave it. Beneath it, in the middle line, is a fold of mucous membrane, the frænum, on either side of which are the orifices of Wharton's ducts, and those of the sublingual glands, or ducts of Riviniani, lie in the fossa between the tongue and maxillary bone.

Structure.—The *mucous membrane* is very adherent to the underlying structure, particularly on the dorsum and sides; on the under surface, however, it is less so, there being a cellular layer between it and the sublingual muscles. The mucous membrane is freely supplied with papillæ; the sides and tip with fungiform, almost its entire surface with filiform, and the posterior part of its dorsum with the V-shaped series of the circumvallate. There are a considerable number of glands lying in this tissue, which give rise to the development of that encysted tumor known as *ranula.* An enlargement of the bursa existing between the hyoid attachment of the genio-hyoglossi, by its enlargement and protrusion beneath the tongue, may be mistaken for ranula.

The *muscles* are both intrinsic and extrinsic. The intrinsic or linguales are two symmetrical bundles of muscular fibre, separated from each other by a fibrous septum, occasionally cartilaginous, and these muscular

fibres are arranged into—(1) the lingualis superior, the
fibres of which are disposed obliquely and longitudinally
on the surface of the organ; (2) an inferior longitudinal

FIG. 12.

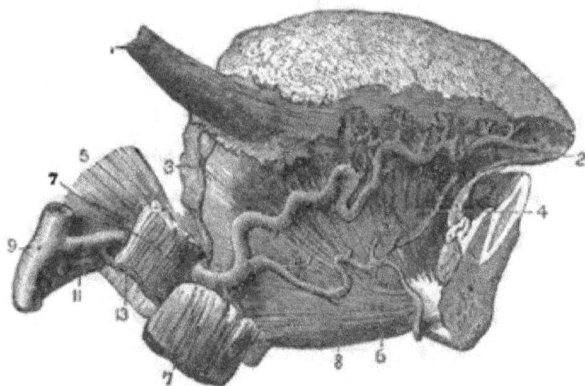

Lingual artery and its branches. 1. Stylo-glossus. 2. Ranine artery. 3. Dorsalis linguæ artery. 4. Genio-hyoglossus muscle. 5. Middle constrictor. 6.
Genio-hyoid muscle. 7, 7. Hyoglossus (cut). 8. Sublingual gland. 9. External
carotid. 11. Lingual artery. 12. Hyoid branch. (HEATH.)

set, passing from the hyoid bone to the apex, and in relation on its under surface with the ranine artery; its
fibres are blended with those of the stylo-glossus; (3)
a transverse set, forming the bulk of the tongue, placed
between the superficial and longitudinal, are attached to
the fibrous septum, and curving outwards are inserted
into the dorsum linguæ and its margin. The fibres interlace with the before-named sets. The existence of the
fibrous septum explains how it is that in acute inflammation of the tongue, or when abscess has formed, the
tumor is frequently unilateral.

The extrinsic muscles are,—the hyo-glossus, genio-
hyo-glossus, stylo-glossus, palato-glossus, and some few
fibres of the superior constrictor.

Arteries.—The lingual artery at the anterior edge of

the hyo-glossi muscles divides into the sublingual and ranine : of these the ranine is the most important; it lies on the under surface of the tongue, external to the genio-hyo-glossi, and on the inner side of the hyo-glossi, stylo-glossi, and sublingual gland. It enters the organ at its base, and runs forwards towards the tip, and in the mouth it lies to the side of the frænum, and is here covered only by mucous membrane ; thus, in dividing this membrane for *tongue-tie*, the scissors should be directed *downwards* and *backwards*.

The ranine artery is generally accompanied by *two* ranine veins, which terminate in the internal and external jugular and the facial veins (*vide* Lingual Artery).

The nerves are very numerous, and are derived from— (1) the hypo-glossal or ninth, which is supplied to the extrinsic muscles only—it is the *motor* nerve of the tongue ; (2) the gustatory branch of the inferior maxillary division of the fifth, supplies the sides and tip—a nerve of *special sense :* these two nerves freely inosculate ; (3) the glosso-pharyngeal, which supplies the circumvallate papillæ at its base—also a nerve of *special* sensation ; (4) the facial, supplying the linguales, by means of the chorda tympani ; (5) the vagus, sending a few filaments to its base ; and (6) the sympathetic—the *vaso-motor* nerve, accompanying the lingual artery.

The base of the tongue is in relation with the epiglottis, which curves forwards towards it during respiration ; but during deglutition it is drawn backwards and downwards, thus covering the aperture of the larynx and preventing food from passing into it (*vide* Larynx).

The tumors met with beneath the tongue are—ranula, salivary concretions, fatty, and bursal.

SURGICAL ANATOMY OF THE PHARYNX.

The pharynx presents for examination four walls,—an anterior, a posterior, and two lateral.

The *anterior*, very oblique behind and below, contains from above downwards, the posterior nares, the velum pendulum palati, the posterior pillars of the fauces, base of tongue, epiglottis, the glosso-epiglottic folds, and larynx.

The *posterior* wall is in relation with the cervical vertebræ, being separated from them by the recti antici and longi colli muscles. Between the muscular coat of the pharynx, which is formed by the constrictors, is a quantity of loose cellular tissue, in which *retro-pharyngeal* abscesses form, frequently from disease of the cervical vertebræ. These abscesses push the pharynx forward against the posterior nares if high up, and if lower down, by pressing upon the larynx interfere with respiration and speech. Occasionally these abscesses point at the side of the neck in front of the sterno-cleido-mastoid (*vide* Fasciæ of Neck).

The *lateral* walls, also formed by the constrictors, are in relation with the sympathetic, glosso-pharyngeal, spinal accessory, vagus, hypo-glossal nerves, internal carotid artery, and internal jugular vein. In front of the vessels and nerves, the pterygo-maxillary region, pterygoid plates of the sphenoid, deep portion of the parotid, and the lateral portion of the submaxillary region are in immediate relation with them. This close relation to the parotid and internal maxillary regions, explains the pointing of abscesses forming in these spaces at the sides of the pharynx.

It is hardly necessary to remind the student that in

passing a bougie, stomach-pump tube, or probang, care must be taken to apply it firmly to the posterior wall of the pharynx to avoid entering the trachea. In making an examination of the pharynx for the detection and removal of foreign bodies, it will be noticed that far more of its cavity can be reached with the finger than might be supposed, and instruments must be used very carefully and sparingly in searching for them; moreover, it must be remembered that the walls of the pharynx are in close contact when not transmitting food.

CHAPTER II.

SURGICAL ANATOMY OF THE NECK.

THE region of the neck is that portion of the body which is contained between the occipital bone above, and the superior aperture of the thorax below—namely, the first two ribs, laterally; the manubrium sterni, anteriorly; and the first dorsal vertebra, posteriorly; and for the sake of simplicity will be divided as follows:

An *anterior,* including the submaxillary or suprahyoid region; a *lateral,* the sterno-mastoid or carotid, the supra-clavicular, the occipital; and a *posterior,* including the nape of the neck,—a method which seems advisable as being a natural one, and agreeable to its external conformation.

Surface Markings.—The development of the neck varies in individuals, both with regard to age and to sex: round and smooth in females and children; muscular, and with all its prominences well marked, in adult males.

As far as its normal length is concerned, in adults it is tolerably constant, the difference in certain persons being rather apparent than real, resulting from some peculiarity in the conformation of the shoulders, &c. Its breadth is variable.

In the natural position, that is, when the base of the skull is parallel to the ground, the *markings* of chief in-

terest to the surgeon are those of the hyoid and laryngeal apparatus and the sterno-mastoids, and the hollows
between and behind these muscles.   The first prominence
below the chin is the pomum Adami, which is far more
prominent in males than in females, and becomes developed at puberty; nearly a finger's breadth above this
can be felt the hyoid bone, with the anterior belly of the
digastric muscle sweeping upward towards the chin.
Immediately below the thyroid cartilages, in the median
line, is a depression, indicating the position of the crico-
thyroid membrane; next, the body of the cricoid cartilage itself; below this, the upper rings of the trachea
may be distinguished, and at about the third ring the
isthmus of the thyroid body can generally be made out,
more particularly in women.   The position of the laryngeal apparatus during swallowing should be noticed, as
it is drawn upwards at the commencement of the act,
returning to its normal position on its completion, the
thyroid body being carried with it.   This fact is of great
value in the diagnosis of tumors in the region of the
trachea or carotid vessels.   In children the trachea is
more deeply placed, very small, and movable (*vide*
Trachea).

The anterior and posterior borders of the *sterno-cleido-
mastoidei* are very evident, even when these muscles are
at rest, from the prominent mastoid processes to their
inferior attachments—the sternal of which is fusiform
and cordlike, and the clavicular, flat and ribbon-shaped,
lying posterior to the former, and variable in its extent
along the clavicle.   The point of divergence of these
two sets of fibres is generally well seen, more especially,
however, when the muscle is in action, as in rotation of
the head from side to side.   These points will be here-

after seen to be important landmarks to the surgeon in
the operative surgery of the neck. Between the angle
of the jaw and the mesial line is the protrusion of the
submaxillary gland, the difference in the position of
which during flexion or extension of the neck, should be
carefully noted. Above the **clavicle**, and between the
prominences **of the** sterno-mastoid and trapezius, is **a**
hollow, the subclavian, or *supra-clavicular fossa*, at the
lower and **internal part of** which the posterior belly of
the omo-hyoid crops up from behind the clavicle. This
muscle lies much more hidden by the sterno-cleido-mastoid **and clavicle than** usually represented in plates of
the posterior triangle. **It** is seen in action after swallowing, during the depression of the **hyoid** bone, **and**
during deep inspiration. The subclavian artery beats
at the bottom of this hollow, and is here readily com-
pressed against the first rib (*vide* Surgical Anatomy of
Subclavian Triangle). The lateral contour of the neck
is **completed, behind, by** the sweep of the trapezius from
the occiput to the tip of the shoulder. Posteriorly the
neck presents a median depression, on either side of
which is seen the mass of the extensor muscles of the
head, and **lower down the** spinous processes of the seventh
and eighth cervical vertebræ. A collection of lymphatic
ganglia is usually very apparent in this region. The
external jugular veins are **seen** on **the** lateral aspect of
the neck, crossing **the** sterno-mastoid obliquely, from
before backwards, at about its middle, **and** passing into
the hollow behind **it.** The more detailed account **of**
these superficial markings is attached to the description
of the several regions of the neck.

   Arrangement of the Cervical Fascia.—**The** attachments and connection of the fasciæ of the neck are of

great surgical importance, inasmuch as these aponeurotic sheaths in a great measure control the course taken by diffuse inflammation, collections of pus, blood, and growths; the latter frequently not appearing externally in the neck until some while after they have extended or sent processes along or amongst them.

For convenience of examination and simplicity of description it may be divided into two layers,—a superficial, and a deep. The *superficial* layer is usually traced from behind, where it commences as a very thin lamina attached to the spinous processes of the cervical vertebræ, superior curved line of the occipital bone, and ligamentum nuchæ; and passing forwards, getting denser as it proceeds, it incloses the trapezius, and, forming sheaths for the posterior muscles of the neck, extends over the posterior triangular space, and arriving at the posterior border of the sterno-cleido-mastoid, forms a sheath for it; and part of it, which constitutes the anterior portion of this sheath, is attached to the lower border of the body and angle of the lower jaw and zygoma, after having covered in anteriorly the parotid gland and masseter muscle; below it is attached to the anterior part of the clavicle and manubrium sterni, and is perforated by the external jugular vein and cutaneous nerves. The *deeper layer* has an attachment to the tubercles of the transverse processes of the cervical vertebræ, and incloses the scaleni muscles, forming the prevertebral aponeurosis, which sends processes over the cords of the cervical and brachial plexuses and subclavian vessels. The lax cellular tissue lying between the prevertebral aponeurosis and the pharyngeal muscles is the seat of retro-pharyngeal abscesses, which point either into the pharynx or, guided by fascia, behind the

carotid vessels (*vide* Pharynx). Passing from the under
surface of the sterno-mastoid towards the middle line,
it is attached to the hyoid bone, forming an aponeurotic
loop on its upper surface, through which runs the tendon
of the digastric; and extending downwards it forms the
sheaths of the sterno-hyoid and sterno-thyroid muscles.
The lamina forming the posterior portion of the sheath
of the sterno-mastoid is attached above to the angle of
the jaw and to the base of the styloid process behind,
and to the inner side of the parotid gland, forming a
septum between it and the submaxillary gland, the *stylo-
maxillary ligament.* The lower portion of this lamina
forms the sheath of the carotid vessels, which sheath is
divided by septa inclosing internally the carotid artery,
externally the internal jugular vein, and posteriorly the
vagus nerve; traced downwards and outwards it is found
to inclose the omo-hyoid muscle, binding it down to the
clavicle, and inclosing the subclavius muscle, passes
beneath it into the axilla, to be continuous with its
fasciæ (*vide* Axilla). At the root of the neck this fascia
is easily demonstrated to be continuous with the pericardium.

That portion of the fascia which is attached to the
hyoid bone above, and to the clavicle and sternum below, has been supposed to have some influence on respiration, inclosing in its reflexions the depressors of the
hyoid bone, and sending processes around the great
venous trunks. The omo-hyoidei in particular by their
contraction tighten it, and being made tense, the calibre
of the veins is increased, and as these muscles only contract during *inspiration*, the dilatation of the veins coin-
cides with the dilatation of the thorax, thereby urging
the blood towards the heart. This, moreover, explains

how readily air may pass into the right side of the heart, should any one of the larger veins, or some branch close to the trunk, be divided.

These several layers of fascia are attached both to the margins of the superior aperture of the thorax and to those structures which pass upwards or downwards through it; and as they are connected together by transverse septa, they constitute a species of diaphragm between the cervical and thoracic regions.

SURGICAL ANATOMY OF THE SUBMAXILLARY REGION.

This region is bounded *above* by the body of the lower jaw, and the floor of the mouth; *below*, by the hyoid bone; *externally,* by the anterior margin of the sterno-cleido-mastoid muscle. Its surface-markings have been already described (*vide* Neck).

Dissection.—On reflecting the integument, immediately beneath it, is the subcutaneous cellular tissue, and the fibres of the platysma myoides muscle passing obliquely from the jaw towards the chest and shoulder, the anterior border of which is free, and separated from its fellow by a cellular interval. Beneath this muscle is a very lax cellular tissue, and on turning it up along the body of the jaw, are seen the nerves supplying it, derived from the facial and upper cervical. A portion of the superior layer of the deep cervical fascia is next met with, attached along the jaw covering in the submaxillary gland, and forming the anterior portion of its capsule; it is continuous externally with the sheath of the sterno-cleido-mastoid, and with the fascia covering the parotid anteriorly. A great many lymphatics lie

either upon or beneath the capsule of this gland, and
can be readily felt beneath the integument. Beneath
this aponeurosis, and inclosed by that portion of it
which is attached to the cornu of the hyoid bone, are
the submaxillary gland, the stylo-hyoid and digastric
muscles. The fibres of the stylo-hyoid are seen inclosing the tendon of the digastric and inserted into the
hyoid bone, just before that muscle passes through the
loop derived from the deep cervical fascia. The muscular fibres of the posterior belly of the digastric are
superior to those of the stylo-hyoid, and after being reflected from the hyoid bone, spread out into a large
muscular mass, having many tendinous intersections
(frequently interlacing with the opposite muscle), to be
inserted into the digastric fossa of the inferior maxilla.
Lying in the interval between the two bellies of the
digastric, and overlapped by its posterior belly, and by
the body of the jaw, is the *submaxillary gland*, inclosed
in its capsule, the posterior portion of which is continuous with the stylo-maxillary ligament, which separates
it from the parotid gland, and superior to it is the facial
artery and vein; the submental branches of these vessels,
with the nerve to the anterior belly of the digastric, pass
forwards towards the symphysis under cover of the body
of the jaw. The bulk of the gland being pulled upwards from the fossa in which it is lodged, it will be
observed that a portion of it passes beneath the mylo-
hyoid muscle, upon which the greater part of it rests.

Parts beneath the Mylo-hyoid.—The mylo-hyoid muscle should now be divided and reflected, when from
before backwards will be seen the genio-hyoid anteriorly,
and deeper down the genio-hyo-glossus, along the outer
side of which lie the ranine vessels, and external to it

the deep portion of the submaxillary and the sublingual glands, and the mucous lining of the floor of the mouth; posteriorly, the hyo-glossus muscle, extending from the cornu of the hyoid bone to the side of the tongue, and upon this muscle from below upwards lie—(1) the ninth or hypoglossal nerve; (2) Wharton's duct; (3) the gustatory nerve with the submaxillary ganglion and chorda tympani, and the inosculating branches of these two nerves. Upon the cornu of the hyoid bone is seen the trunk of the lingual artery just before it disappears behind the hyo-glossus muscle.

Arteries.—The *facial* artery in this region passes beneath the posterior belly of the digastric and stylohyoid and submaxillary gland, and after giving off branches to the gland, and the submental, it becomes subcutaneous, about an inch in front of the angle of the jaw.

The *lingual* artery is deeper in its course and distribution, and is directed towards the hyo-glossus, and is at first covered by the skin, platysma, and fascia, and rests on the middle constrictor; after passing over the cornu of the hyoid bone, it is crossed by the ninth nerve, the digastric and stylo-hyoid muscles forming an arch over it. It is next covered by the hyo-glossus, and lies on the superior constrictor and genio-hyo-glossus muscles.

The *branches* given off from it are,—the hyoid, which runs along the upper border of the hyoid bone; the dorsalis linguæ, supplying the dorsum of the tongue, tonsil, and soft palate; the sublingual, supplying its substance, and the ranine.

Ligature of Lingual Artery.—To place a ligature upon the lingual artery, that portion of its course where it lies upon the great cornu of the hyoid bone is selected, im-

mediately before it passes behind the outer border of the
hyo-glossus muscle, as it is there most accessible; and to
reach it, an incision is to be made about a finger's breadth
below the body of the jaw through the integument and
aponeurosis, and the gland is to be lifted upwards. After
the posterior portion of the capsule of the gland has been
divided, the combined lingual and facial veins are seen
passing obliquely backwards, and deeper, the hypoglossal nerve; at the angle where this nerve meets the
tendon of the digastric, lies the artery, taking a curve
downwards towards the hyoid bone. Occasionally the
vessel pierces the hyo-glossus muscle, or this muscle extends farther back than usual, in which case its fibres
must be divided. On the dead body the vessel appears
to be tolerably near the surface, but during life the fascia
and integuments are so on the stretch and so attached
to the salient parts of the region, that when the vessel
is exposed, it is actually very deep. The operation is a
difficult one, the vessel being only supported by the loose
wall of the pharynx, which runs considerable danger of
being wounded, so that the only sure firm guide to it is
the posterior cornu of the hyoid bone. The hyoid bone
may be with advantage drawn forwards into the wound
and steadied with a hook (*vide* Region of Tongue).

Veins.—There is often a considerable plexus of veins
in this region, the most important being the facial and
the lingual; the facial leaves its artery and passes upon
the fascia in front of the submaxillary gland, whilst
the lingual vein is separated from its artery by the hyo-
glossus muscle. Very often these veins form a common
trunk, lying superficial to the hypoglossal nerve, before
entering the jugular vein.

Nerves.—The superficial nerves have been already

referred to; the hypoglossal enters the region superficial to the external carotid and below the stylo-hyoid and digastric muscles, and passes upwards over the cornu of the hyoid bone between the glands and the hyo-glossus muscles, covered in by the mylo-hyoid; it loops with the gustatory and is seen distributed to the extrinsic muscles of the tongue. The gustatory nerve, with the chorda tympani, lies beneath the gland and passes to the mucous membrane of the sides and tip of the tongue, upon its deeper portion.

The above relations are those borne by the different structures in this region, when the head is in its natural position, but when the parts are the seat of operation, the head is thrown back and the contents of the space put on the stretch; by so doing the position of the sub-maxillary gland and the facial vessels are considerably altered by the cavity which naturally exists between the lower jaw and the mylo-hyoid muscle becoming flattened, thereby causing a protrusion of the structures between them.

SURGICAL ANATOMY OF THE INFRA-HYOID REGION.

The boundaries of this region are,—*above*, the hyoid bone and the base of the tongue; *laterally*, the sterno-mastoid muscles (and carotid vessels); and *below*, the upper border of the manubrium sterni, or interclavicu-lar notch; *posteriorly*, the cervical vertebræ, covered by the longi colli muscles. This region is surgically im-portant as containing the larynx and trachea, the cervi-cal portion of œsophagus, and the thyroid body, with their respective vessels and nerves. Its surface mark-ings have been already described (*vide* Neck).

Dissection.—On the removal of the integument, the subcutaneous cellular tissue is first met with, and the anterior portion of the platysma, which is usually unconnected with its fellow of the opposite side, there being a cellular interval between them, well marked in the necks of old people when the fat is absorbed, causing the "dewlap" appearance characteristic of age. Beneath these muscular fibres is a thin layer of cellular tissue, which allows of the free movements of the integument and platysma over the underlying aponeurotic sheaths of the muscles. Along the anterior border of the sterno-mastoid is the anterior jugular vein, which, however, is at times absent. The sterno-hyoid and omo-hyoid muscles themselves form the next layer, in their aponeurotic sheaths, and immediately below them lie the sterno-thyroid and thyro-hyoid muscles. The sterno-hyoid muscles pass somewhat obliquely, so that the interspace between their internal free borders is rather wider towards the sternum than at the hyoid bone; whilst the inner margins of the sterno-thyroids are slightly oblique in the opposite direction.

The nerves supplying these muscles are seen generally on their posterior borders, or ramifying upon them. Beneath the muscles just named, in the mesial line, from above downwards, are met with—(1) the hyoid bone; (2) the thyro-hyoid membrane; (3) the thyroid cartilage; (4) the crico-thyroid membrane; (5) the cricoid cartilage, partly covered by the crico-thyroid muscle; (6) the first ring or two of the trachea; (7) the thyroid body and its isthmus extending between the lobes; (8) the trachea (passing backwards and downwards towards the thorax), upon which lies a plexus of veins, the inferior and middle thyroid, passing downwards from the **thyroid body ; and** the thyroidea ima artery, when it exists, is generally a branch of the innominata. On the right side of the neck, the common carotid artery crosses the lower portion of the trachea obliquely, but subsequently lies parallel to it ; on the left side the common carotid is deeper **than on the** right **and lies** along the trachea. It must **be borne in** mind **that** neither of these vessels is in *actual* contact with the trachea. Behind the trachea is the œsophagus, **which commences opposite the fifth cervical vertebra and cri-** coid cartilage, and passes to its left side—a circumstance which is taken advantage of for the performance of **the** operation of œsophagotomy. On either side is the common carotid artery, the nearer being the left, owing to the projection of the œsophagus towards that side ; **and** crossing it obliquely are the superior and **inferior thy-** roid arteries. The recurrent laryngeal **nerve** lies **in the** interspace between the borders of the trachea and œsoph- agus, and passes below the inferior constrictor and into the larynx through the crico-thyroid membrane. The œsophagus **is separated from the** cervical spine and the muscles covering it—viz., **the** recti antici, **longi colli,** and that portion of **the deep** cervical fascia **which is** applied to their anterior surface, by a layer of **very lax** cellular tissue, which permits of the constant **gliding of** the œsophagus and trachea in their several movements, and is, as before mentioned, frequently the seat of ab- scesses. For greater convenience of reference, two tri- angular spaces, termed by Velpeau the *omo-hyoid* and the *omo-tracheal,* **may be** noted ; the former, bounded by the hyoid bone above, sterno-mastoid externally, and the omo-hyoid internally, contains the superior thyroid artery, superior laryngeal **nerve, a portion of**

7

the middle and inferior **constrictors, alæ of** thyroid cartilage, and thyro-hyoid membrane ; the latter is bounded
above and externally by the omo-hyoid, below and externally by the sterno-mastoid, **and** internally by the
middle line of the neck, and contains the sterno-hyoid
and thyroid muscles, a lobe of the thyroid body, the
superior and inferior thyroid arteries, descendens and
communicans noni nerves, sides of cricoid cartilage and
trachea, the **recurrent laryngeal nerves,** and, on the left
side, the œsophagus.

The superior thyroid artery arises from the external
carotid artery, opposite the greater cornu of the hyoid
bone, and at first lies superficially, in a space bounded
by the sterno-mastoid, digastric, and omo-hyoid muscles.   It then passes upwards and inwards, and arches
down to the upper part of the lobe of the thyroid body,
lying below the omo-hyoid, sterno-hyoid, and sterno-
thyroid muscles, having behind it the superior laryngeal nerve.   Its crico-thyroid branch runs transversely
across the thyro-hyoid membrane, and may be wounded
in laryngotomy.

The inferior thyroid artery, in this region, passes obliquely upwards and inwards, crossing behind the common carotid artery, internal jugular vein, pneumogastric, and sympathetic nerves ; and on the left side it
lies in front of the œsophagus and behind the thoracic
duct, and enters the lower part of the lobe of the thyroid body.   These vessels very freely anastomose with
each other, and with those on the opposite side.   The
superior laryngeal nerve lies deep down in this region,
passing behind the external and internal carotids, and
divides into two branches—an external, supplied to the
crico-thyroid muscle, and a deep one, penetrating the

thyro-hyoid membrane and distributed to the mucous membrane of the larynx (*vide* Subclavian **Artery**). This region is of great surgical importance; as well as being the usual seat of injuries inflicted suicidally or homicidally, the operations of laryngotomy, cricotomy, tracheotomy, and œsophagotomy are performed therein.

Non-surgical wounds, whether suicidal or otherwise, are invariably made *across*, and, as a rule, the main vascular trunks escape, unless the attempt be very determined, for the head is thrown back and these main trunks recede, on the structures beneath the sterno-mastoid being put on the stretch. The usual situation selected for such attempts is the thyro-hyoid space, and the incisions frequently divide the larynx and some branches of the superior thyroid artery, and not unfrequently cut through the base of the tongue and epiglottis.

All openings made *surgically* into the air-passage are made in the *middle* line, for the very important reason that the muscular coverings of the trachea do not unite in the mesial line, but merely approximate, leaving a cellular interval, through which the windpipe is reached. Unless the middle line be adhered to, although the trachea may be opened, great difficulty will probably be experienced in the introduction of the tube, as it will have a tendency to slide between the muscles and the trachea, and miss the opening made in it. Again, supposing no difficulty to arise of this nature, the muscle is so much wounded as to be seriously impaired, and may become united to the integument after the tube is removed.

There are three methods by which the windpipe may be opened, viz. :

Laryngotomy, in which the crico-thyroid membrane is divided ; this is the readiest method of admitting air, the only difficulty which might arise being the hemorrhage from the crico-thyroid arteries, which run across this space. The incision in the membrane, which is made horizontally, must not be so wide as to injure the crico-thyroid muscles.

Tracheotomy is the operation in which the rings of the trachea are divided either above or below the isthmus of the thyroid body. It must be borne in mind that, although the trachea is very superficial above, it recedes, and is very deep below, and, just above the sternum is generally at least an inch from the surface, its depth varying according to the amount of fat or muscle in the individual, or to the incurvation of the cervical vertebræ. It is crossed at about its second or third rings by the isthmus of the thyroid body, and below the isthmus, on the lower part of the trachea, lie the middle and inferior thyroid veins, which are generally greatly engorged, on account of the asphyxia necessitating the operation. Occasionally the rings are ossified in old persons, and may cause trouble.

In *children,* owing to the shortness of the neck, and the depth, small calibre, and mobility of the trachea, the operation is a difficult one. Again, the plexus of veins in connection with the thyroid body and the closeness of the carotids, render it one demanding the greatest care. The innominata is proportionably higher up in the neck than in the adult, on account of its obliquity. In children under two years of age, and in some instances far older, the thymus gland may give great trouble by bulging up into the wound, and so obstructing the operator's view of the parts.

Cricotomy consists in dividing the cricoid cartilage ; but it is an operation rarely resorted to.

Foreign bodies in the trachea are naturally directed towards the *right bronchus*, because it is wider than the left. The septum at the bottom of the trachea, which separates the bronchi, occupies the left of the median line.

The right bronchus is shorter and more horizontal than the left, being about one inch long. The left is about two inches in length, and is directed more obliquely than the right.

SURGICAL ANATOMY OF THE STERNO-MASTOID RE-GION, OR REGION OF THE CAROTID ARTERIES ABOVE THE STERNO-CLAVICULAR ARTICULA-TION.

The boundaries of this region may be sufficiently stated as being those occupied by the sterno-cleido-mastoid muscle itself. This muscle forms an oblique rectangular eminence, and is attached above to the mastoid process and the superior curved line of the occipital bone, its broad tendinous *insertion* being blended with the fibres of origin of the trapezius ; passing downwards and forwards, its muscular fibres become divided, the anterior set collecting themselves into a round fusiform bundle, to be attached to the manubrium sterni, and a posterior bundle, a flattened riband-like band, separated from the former by a cellular interval, and attached for a variable distance along the inner and upper aspect of the clavicle. (Occasionally these clavicular fibres pass along the whole inner two-thirds of the clavicle, forming a muscular layer, almost covering in the posterior triangular space.) This interval in the disposition of the muscular fibres is

of great surgical importance. The anterior border is the
more prominent, and is rounder than the posterior, which
becomes lost in the general surface of the posterior part
of the neck. In most works on descriptive anatomy it
is not sufficiently enforced that this muscle completely
covers in the common, internal, and external carotids,
and that while the muscle, its integuments, and the
fascial coverings are intact, it may surgically be considered to extend forwards as far as the angle of the jaw.

Its pulsations are in reality felt beneath its own border,
or immediately beneath its sternal and clavicular attachments.[1]

Structures superficial to the Sterno-mastoid.—On reflecting the skin and cellular tissue the first structure
met with is the platysma, passing obliquely backwards
from the jaw to the shoulder, and on dissecting off this
layer of muscular tissue from above downwards, the
structures met with are, some filaments of the small occipital nerve, the great auricular nerve, the external
jugular vein, which usually crosses the muscle obliquely
at about its middle, to pass ultimately into the subclavian vein, the transverse superficial cervical nerve, passing forwards to the under surface of the platysma and
integument, and some branches of the descending clavicular nerves. The disposition of the fibres of the platysma, as far as regards the direction of the external
jugular vein, is important, as in venesection it is necessary to cut *across* them, and not in their continuity,
otherwise the wound would close from muscular con-

---

[1] According to Richet it is impossible to puncture the common
carotid from the side of the neck without perforating the sterno-
mastoid in the *undissected* subject, a statement I have taken care
to verify.

traction. That portion of the cervical aponeurosis which forms the anterior layer of the sheath of the muscle is next seen, attached above to the angle of the jaw (almost appearing to divert the anterior border of the muscle

FIG. 13.

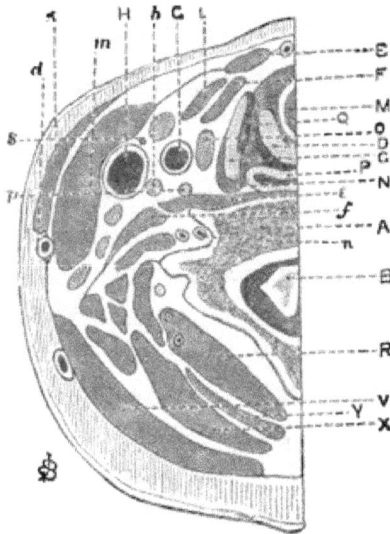

Diagram of the parts seen in a horizontal section through the sixth cervical vertebra. A. Body of sixth cervical vertebra. B. Spinal cord. C. Thyroid cartilage. D. Cricoid cartilage. E. Sterno-hyoid muscle. F. Omo-hyoid. G. Common carotid artery. H. Internal jugular vein. K. Platysma. L. Sterno-thyroid. M. Opening of larynx. N. Inferior constrictor. O. Summit of lateral lobe of thyroid body. P. Œsophagus. Q. Thyro-arytenoid muscle. R. Spinalis colli. V. Trapezius. X. Splenius. Y. Complexus. d. External jugular vein. b. Vagus nerve. e. Longus colli. f. Scalenus anticus. m. Sterno-cleido-mastoid. n. Vertebral vessels. p. Sympathetic. s. Descendens noni.

from the straight line), and below to the clavicle, and to the corresponding facial sheath of the opposite muscle; and after completely inclosing the muscle at its posterior border, it becomes continuous with the aponeurosis of the neck.

The sterno-cleido-mastoid is perforated on its under surface, near the centre, by the spinal accessory nerve,

which entering it obliquely, after inosculating in its substance with the second and third cervical nerves, passes
out behind its posterior border, and crosses the posterior
triangular space. The muscle is readily seen in action,
on rotating the head, or on bowing it upon the thorax,
when *both* muscles are used.

Parts beneath the Sterno-cleido-mastoid.—Immediately
behind the muscular fibres is the posterior layer of the
sheath ; and between it and the muscle lie a considerable
number of lymphatics, and some twigs of the sterno-
mastoid branches of the superior thyroid artery.

The most convenient method of grouping the structures
which lie beneath the muscle is to divide it into three
portions, making the crossing of the omo-hyoid to separate the middle and lower, and a line drawn backwards
from the angle of the jaw to separate its middle and
upper. Beneath the posterior layer of the sheath in the
upper third, from above downwards, the structures met
with are,—the insertion of the splenius capitis, beneath
which are the attachments of the posterior belly of the
digastric and of the trachelo-mastoid, the posterior auricular and occipital vessels, the external border of the complexus, while still deeper are the attachments to the atlas,
of the obliqui and rectus lateralis muscles, the vertebral
artery, and suboccipital nerve. In the middle third,
passing obliquely into the under surface of the muscle,
is the spinal accessory nerve, which unites with some
filaments of the cervical plexus ; next appear a chain
of lymphatic glands (*glandulæ concatenatæ*), and the
branches of origin of the superficial cervical plexus, the
descendens and communicans noni nerves. Beneath
them lie the common, internal and external carotid arteries and jugular vein, the hypoglossal nerve, the vagus,

the superior cervical ganglion of the sympathetic, and deeper down upon the spinal column the attachments of the rectus anticus major, scaleni, levator anguli scapulæ, and splenius colli muscles.   Next is the crossing of the omo-hyoid, and in the inferior third from within outward, are the outer borders of the sterno-hyoid and thyroid muscles, covered by their aponeuroses, with the nerves supplying them.   Posteriorly and externally are the scaleni and the cervical plexus, the phrenic branch of which lies on the anterior scalenus, behind which is the third part of the subclavian artery, having the ascending cervical artery lying parallel and internal to it; a quantity of lymphatic ganglia, and, inclosed in their proper sheath, lying obliquely, in the middle of this space, the common carotid artery and the internal jugular vein which joins the subclavian vein below; into the junction of which pass, on the right side, the common lymphatic trunk, and on the left the thoracic duct with its tributaries.   Posteriorly, are the vagus and the recurrent laryngeal nerves, and closer down on the spine the cord of the sympathetic and the middle cervical ganglion, lying on the prevertebral aponeurosis.   Posterior to the carotid vessels and below, is that portion of the *subclavian artery* which lies internal to the anterior scalenus.   This vessel, on the right side, arises from the innominata, and lies immediately behind the inferior angle of the divergence of the sternal and cleidal origins of the cleido-mastoid, and is separated from the sterno-clavicular articulation and origins of the sterno-hyoid and thyroid muscles, by the junction of the internal jugular and subclavian veins.   The vagus and phrenic nerves lie in front of it, with numerous branches of the sympathetic; whilst embracing it, and passing behind it, is the recurrent

laryngeal nerve. Behind it is the transverse process of
the seventh cervical vertebra, and internally the common
carotid itself; below and externally, this portion of the
subclavian artery is in relation with the pleura.

The branches of the *subclavian artery* being normally
derived from the first part of its course, it follows that
on the right side these branches lie beneath the clavicular portion of the sterno-cleido-mastoid muscle; and this
fact, added to the mechanical difficulty of reaching it,
forms a serious obstacle to success on placing a ligature
upon it in this situation. In the event of the operation
being undertaken, it should be tied as near the vertebral
as possible, so that a coagulum may be formed between
this point and the origin of the trunk.

On the *left* side the recurrent laryngeal is not in relation with the subclavian artery in the neck (*vide* Subclavian Artery).

The sheath of the carotid vessels is derived from the deep
cervical fascia, and is divided by a septum into three
compartments—the inner containing the artery, the external the vein, and the posterior the vagus nerve;
whilst either on it, or sometimes in it, is the loop formed
by the descendens and communicantes noni. Beneath
the sterno-mastoid, the relations and course of the right
and left common carotids are precisely similar. Their
course is represented by a line drawn from the sterno-clavicular articulation, to the external aspect of the upper
border of the thyroid cartilage, at which level generally
it divides into external and internal carotid.

The left common carotid in the neck is a little deeper
and rather longer than the right.

*Relations of the Cervical Portion of the Common Caro-
tid.*—*In front:* Integument, platysma, sternal origin of

sterno-mastoid, sterno-hyoid, sterno-thyroid, omo-hyoid, descendens noni, artery to sterno-mastoid, middle and superior thyroid, lingual and facial, and anterior jugular veins.

Externally: Internal jugular vein, vagus nerve, and lymphatics.

Internally: Trachea, thyroid body, recurrent laryngeal nerve, inferior thyroid artery, larynx, and pharynx.

Behind: Prevertebral muscles, sympathetic, inferior thyroid artery, and recurrent laryngeal nerve.

The external carotid artery is given off from the main trunk, usually opposite the upper border of the thyroid cartilage; it is at first a little *internal* to, and in front of, the *internal* carotid, and passes upwards and forwards, and afterwards a little backwards, towards the angle of the jaw. Up to the level of a line drawn from the mastoid process to the hyoid bone the artery is superficial, but at this point it gets deeper, being crossed by the ninth nerve, the posterior belly of the digastric and stylo-hyoid muscles and a plexus of veins; it then enters the lower border of the parotid gland (*vide* Parotid Region).

The branches of the external carotid are usually given off in the following order: (1) superior thyroid; (2) lingual; (3) facial; (4) occipital; (5) posterior auricular; (6) ascending pharyngeal; terminating in the temporal and internal maxillary.

Relations of External Carotid.—*In front:* Integument, platysma, and fasciæ, sterno-mastoid, hypoglossal nerve, lingual and facial veins, posterior belly of digastric and stylo-hyoid muscles, and parotid gland.

Behind: Superior laryngeal nerve, styloid process, stylo-glossus and stylo-pharyngeus muscles, glosso-pharyngeal nerve, and that portion of the parotid gland which separates it from the internal carotid.

Internally: Hyoid bone, pharynx, parotid, ramus of jaw, and stylo-maxillary ligament.

Internal Carotid Artery (*cervical portion*).—Arises opposite the upper border of the thyroid cartilage, and is at first superficial and external to the external carotid, until the crossing of the digastric, where it becomes deeper and lies beneath the external carotid. Usually it gives off no branches in the neck, and is larger than the external in the child, but of much the same calibre in the adult.

Relations of the Cervical Portion of the Internal Carotid. —*In front:* Integument and platysma, sterno-mastoid, parotid, hypoglossal nerve, styloid process, stylo-glossus, and stylo-pharyngeus muscles, glosso-pharyngeal nerve and its branches.

Externally: Internal jugular vein, and vagus nerve.

Internally: Pharynx, ascending pharyngeal artery.

Behind: Rectus anticus major muscle, sympathetic, and superior laryngeal nerves.

This region is the seat of most important operations— viz., ligature of the common carotid arteries or their branches, of the subclavian in the first part of its course, of the innominata, of œsophagotomy, the removal of tumors, and opening of abscesses.

Ligature of the Common Carotid Artery.—In applying a ligature to the common carotid, that portion of it which lies either immediately above or immediately below the crossing of the omo-hyoid should be selected, as the vessel is there most easily reached.

Above the Omo-hyoid.—The incision to be made varies in length, according to the nature of the case, and the depth of the superjacent structures, but is usually one about three inches in length, along the anterior border

FIG. 14.

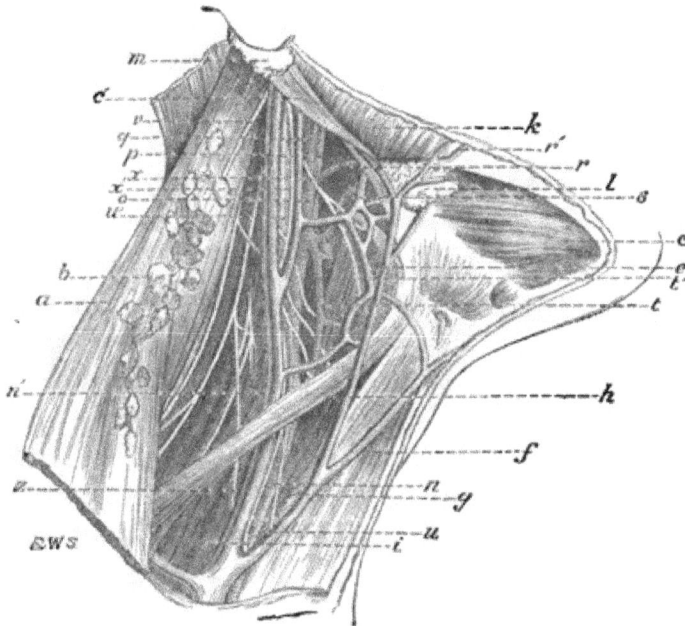

Common carotid artery and its branches. *a.* Sterno-mastoid reflected. *b.* Glandulæ concatenatæ. *c.* Anterior belly of digastric. *c'.* Posterior belly of digastric. *e.* Thyro-hyoid. *f.* Sterno-thyroid. *g.* Sterno-hyoid. *h.* Omo-hyoid. *i.* Anterior scalenus. *k.* Masseter. *l.* Submaxillary gland. *m.* Parotid gland. *n.* Common carotid. *n'.* Internal jugular vein (joined by the anterior jugular and supra-scapular). *o.* External carotid. *p.* Internal carotid. *r.* Facial. *r'.* Facial vein. *s.* Lingual. *t.* Superior thyroid. *t'.* Nerve to thyro-hyoid. *u u'.* Vagus. *v'.* Spinal accessory nerve. *x.* Hypoglossal nerve. *x'.* Descendens noni nerve. *z.* Phrenic nerve.

of the sterno-mastoid, from just below the angle of the jaw to the cricoid cartilage, dividing the integument, superficial fascia, and platysma, and the wound being kept open by retractors; the deep fascia is met with, which is very adherent to the sheath of the vessels; a plexus of veins, and a few small arterial twigs, are often

interspersed between the fascia and the sheath. This
fascia should be cautiously divided on a director, so as
to expose the sheath, upon or beneath which is the descendens noni nerve.

Next a small portion of the sheath is to be pinched
up and "nicked," by holding the blade of the knife
horizontally, immediately over the inner aspect of the
vessel, as far from the vein as possible; and an aneurism
needle is to be passed *from without inwards*, and kept
closely round the artery, so as to avoid wounding the
internal jugular vein or including the pneumogastric
nerve. There is a tough layer of areolar tissue between
the sheath and the artery, which must be gently "teased"
through, by the needle, on being pushed against the
finger nail. The jugular vein may be compressed above
and below during the operation, as it is liable to become
suddenly so distended as to conceal the parts. Should
any difficulty in reaching the vessel be met with, owing
to engorgement of the veins, an important landmark
will be found in the anterior tubercle of the transverse
process of the sixth cervical vertebra, which is behind
and a little internal to the carotid process, and against
this the carotid may be compressed.

Below the crossing of the Omo-hyoid.—Tying the vessel
below the omo-hyoid is much more difficult, owing to its
greater depth, and to the size of the veins: an incision
should be made about three inches in length from the
cricoid cartilage, along the anterior border of the sterno-
cleido-mastoid (which is to be drawn outwards), taking
care to avoid wounding the lower sterno-mastoid artery
and the middle thyroid vein: the fascia covering the
sterno-hyoid and sterno-thyroid muscles is next seen,
and must be cautiously divided, and these muscles

pulled inwards, the sheath being reached (upon which are branches of the loop of the descendens and communicans noni nerves), it is to be opened, and the needle passed *from without inwards.* The inferior thyroid artery and sympathetic and recurrent laryngeal nerves lie immediately behind the vessel in this part of its course. It must be borne in mind that, on the right side of the neck, at its lower part, the internal jugular vein diverges from the artery; but on the left approaches it, and sometimes crosses it, owing to the formation of the innominate veins. As before mentioned, advantage may be taken of the natural interspace between the two heads of origin of the sterno-mastoid to place a ligature on this portion of the vessel, but it is very difficult for the reason just stated (Sédillot). Another method of reaching it is to expose and divide the sternal attachment of the sterno-mastoid, thereby obtaining greater room and corresponding safety (Malgaigne).

Collateral circulation after Ligature of the Common Carotid.—Supposing the vessel to be normal—that is to say, that it gives off no branch before the usual bifurcation, the collateral circulation is very free, and is re-established by vessels both without and within the cranium; thus, the current of blood being arrested in the carotid, the subclavian of the same side becomes dilated, the work outside the skull is thrown upon the inferior thyroid branch of the thyroid axis, the superior thyroid branch of the external carotid, the profunda cervicis of the superior intercostal, and the princeps cervicis of the occipital; the vertebral doing the work of the internal carotid, *within* the skull.

Ligature of the External Carotid Artery.—An incision is to be made similar to that for tying the common carotid above the omo-hyoid, where the vessel is most superficial, and immediately beneath the skin, platysma, and superficial fascia, and a complicated plexus of veins. The posterior belly of the digastric and its attendant muscle, the stylo-hyoid, should be drawn upwards towards the jaw, and the sterno-mastoid outwards; the superior laryngeal branch of the vagus lies usually just behind its short trunk.

The collateral circulation after ligature of the external carotid would be readily maintained by its branches anastomosing so freely on the face with those of the opposite side, and by the terminal branches of the internal carotid (supra-orbital, ethmoidal, palpebral, and nasal) with the facial, and by the profunda and princeps cervicis.

(This operation is not often resorted to in the practice of surgery, as it is preferable to tie the common trunk.)

Tenotomy.—The attachments of the sterno-cleido-mastoid to the clavicle and sternum occasionally require division, *subcutaneously,* for the relief of wryneck or *torticollis,* and considerable caution is requisite in this apparently simple operation, as there is a danger of wounding the external jugular vein as it passes into the internal jugular, or even more important vessels, if it be clumsily or hastily performed.

SURGICAL ANATOMY OF THE SUBCLAVIAN REGION (OR REGION OF THE THIRD PART OF THE SUBCLAVIAN ARTERY).

This region receives its name from the fact of its containing the third portion of the subclavian artery, and on account of its being the most common seat of operation upon that vessel, but from its actual position and natural boundaries it would be more correctly termed *supra-clavicular*. The hollow which exists above and behind the clavicle is almost invariably well marked, even where there is much fat in the neck ; and it is this space, with its numerous contents and varying conformation, which is of such surgical import. It is bounded anteriorly by the posterior border of the sterno-cleido-mastoid ; behind, by the rounded anterior border of the trapezius—these muscles nearly meet above at their cranial attachment, where their aponeuroses are blended ; below by the clavicle, and above by the crossing of the posterior belly of the omo-hyoid, and its floor is formed by the first rib and the muscular structures attached to it.

The pulsation of the subclavian artery can be felt at the bottom of the space as it crosses the first rib, against which it can be readily controlled for any operation about the shoulder or arm. The positions of the scalene muscles, the cords of the brachial plexus, particularly those of the fifth and sixth, and the course of the omo-hyoid, are also felt, and generally to be seen, as eminences beneath the integument. It must be borne in mind that the "triangle" formed by the crossing of the omo-hyoid is a result of *dissection*, and the detachment of its aponeurosis ; no such *regular* interspace existing during life, the inner border of the muscle lying behind the clavicle, and its upper border only being seen whilst in action. The change in the appearance of the hollow immediately above and behind the clavicle is noticeable under certain circumstances ; thus, in *inspiration*, it is considerably deepened, and during *expiration* becomes flatter, when the pulse in the subclavian *vein* is gener-

ally visible. Again, the various movements of the arm
and of the clavicle cause considerable modifications of

FIG. 15.

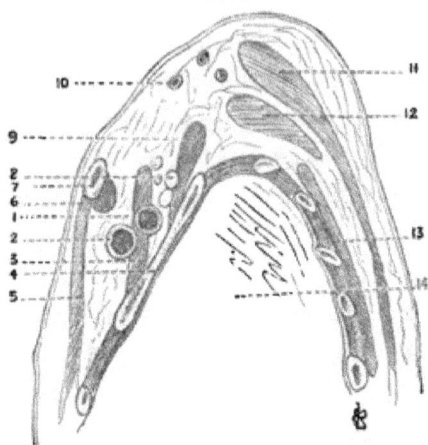

Diagrammatic section through the centre of the right clavicle, showing the
relation of the subclavian vessels in their antero-posterior direction. 1. Subclavian artery. 2 Subclavian vein. 3. Anterior scalenus muscle. 4. First rib. 5. Pectoralis major. 6. Subclavius muscle. 7. Clavicle. 8. Cords of brachial plexus.
9. Scalenus medius. 10. Transversalis colli artery. 11. Trapezius. 12. Levator
anguli scapulæ. 13. Rhomboid. 14. Cavity of thorax.

its form, and the relations of its contents—a circumstance of great importance to the surgeon.

Compression of the subclavian artery is generally necessary in amputations about the upper arm, and very
slight pressure is requisite. The thumb or finger is to be
slipped just behind the posterior border of the sterno-
mastoid, where it is attached to the clavicle, and on slight
pressure being made in a *vertical* direction to the axis of
the body, the pulsation of the vessel is felt; a little further pressure at the pulsating point compresses it against
the first rib, and does not interfere with the circulation
in the subclavian vein. The circulation may be con-

trolled also, in many instances, by drawing the arm backwards, and forcibly depressing the tip of the shoulder.

Dissection.—After removing the skin, superficial fascia, and the posterior fibres of the platysma, with the thin fascia immediately beneath this, the handle of the knife will be generally found sufficient to expose all the important relations. The first structures met with, are the descending superficial branches of the cervical plexus (the acromial and clavicular), and a considerable number of lymphatic glands. Beneath the layer of the cervical fascia which covers in the space is a quantity of loose cellular tissue and fat, in which lie lymphatics and superficial cutaneous and glandular vessels. The upper border of the omo-hyoid, inclosed in its sheath of deep cervical fascia, is next met with, the anterior layer binding down the subclavian vein against the clavicle, which vein here receives the external jugular just behind the attachment of the sterno-cleido-mastoid. Immediately behind the vein, internally, is the anterior scalene muscle; and emerging from behind it, and meeting the vein at an acute angle in a plane superior to it, is the subclavian artery; and in a plane more posterior, but above, are the cords of the brachial plexus. The anterior scalene muscle is inserted into a tubercle (*the scalene tubercle*) on the first rib, which is a guide to the vessel, and in front of, and behind the tubercle are grooves on the rib, in which lie, in the anterior, the subclavian vein, and, in the posterior, the subclavian artery, separated from each other by the muscle. The suprascapular artery, a branch of the subclavian in the first part of its course (normally), with its vein, crosses the main trunk just below the crossing of the omo-hyoid, and lies along the upper border of the clavicle. The

transversalis colli artery, also a branch of the main
trunk in the first part of its course, lies superior to the

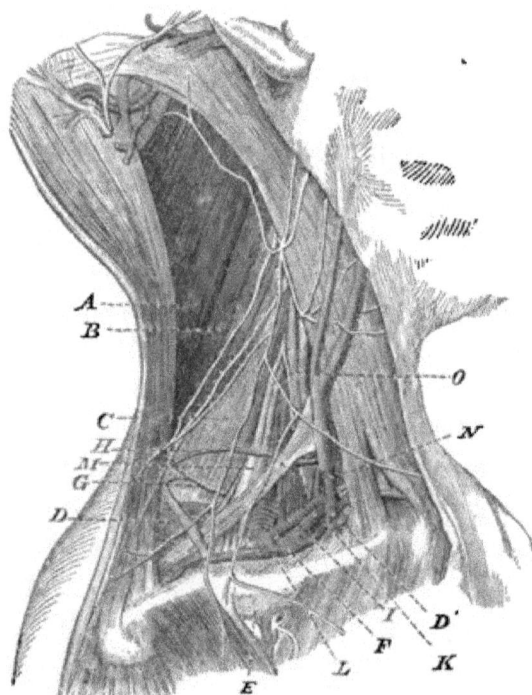

FIG. 16.

Region of the third part of the subclavian artery (the shoulder represented
depressed). A. Splenius. B. Levator anguli. C. Scalenus posticus. D. First
serration of serratus magnus. E. Costo-coracoid membrane and cephalic vein.
F. Subclavian artery. G. Transversalis colli artery (deep). H. A more superficial branch. I. Supra-scapular artery. K. Subclavian vein. L. Supra-scapular
vein. M. Brachial plexus. N. Scalenus anticus. O. Phrenic nerve.

artery, but beneath the cervical plexus it passes towards
the trapezius, to reach the posterior border of the scapula. These vessels are accompanied by veins forming
a plexus, which lies superficial to the artery, and may
cause considerable difficulty in any operation in this
region, especially when engorged.

The relations of the subclavian artery in the third part of its course are :

In front.—Integument, superficial cervical fascia, platysma, external jugular vein, and the venous plexus before mentioned, descending cervical nerves, subclavius muscle, supra-scapular artery, and clavicle.

Above.—Brachial plexus, and posterior belly of omo-hyoid muscle.

Below.—First rib, covered by the first serration of the serratus magnus.

Behind.—The middle scalene muscle.

Ligature of the Subclavian Artery in the third part of its course.—The point selected for placing a ligature upon this vessel, is just where it lies on the first rib, at the bottom of the hollow above described. It is not inclosed in a definite sheath like the carotid, but is bound down by a process of the deep cervical fascia, derived from the aponeurotic investment of the omo-hyoid. The incision is most advantageously made, by drawing the integument down, and cutting upon the clavicle, allowing it to retract afterwards; but this proceeding must be modified by circumstances. When the vessel is reached, just as it emerges from behind the anterior scalene muscle, the needle must be passed round it from *before, backwards ;* and the end of the needle must be made to insinuate itself round the vessel closely, so as to avoid the united cord of the eighth cervical and first dorsal nerves, which lie immediately behind it.

The collateral circulation developed after ligature of the third part of the subclavian artery is as follows :

As the third portion rarely gives off any branches, the blood would pass towards the arm by the supra-scapular and posterior scapular branches of the thyroid axis, anastomosing directly with the dorsalis scapulæ of the axillary, on the dorsum scapulæ. The internal mammary trunk of the subclavian anastomoses with the acromio-thoracic, long thoracic, and subscapular, and the superior intercostal with the superior thoracic.

SURGICAL ANATOMY OF THE INNOMINATA.

Although the innominata, anatomically speaking, belongs to the thorax, yet any surgical proceeding in connection with it would be attempted in the lower part of the carotid region; hence it has been thought advisable to introduce it into the surgical anatomy of the neck.

In the *neck* it, normally, ascends obliquely in the right side to the posterior aspect of the sterno-clavicular articulation, where it divides into the right subclavian and common carotid arteries, and is about an inch, or rather more, in length.

Relations. — From before backwards the structures covering it are—skin, fascia, some fibres of platysma, and descending branches of cervical plexus, sternal origin of sterno-mastoid, a portion of the manubrium sterni, origins of the sterno-hyoid and sterno-thyroid muscles, remains of thymus gland, left innominate, and right inferior thyroid veins, and cardiac branches of vagus.

On its right side lie the right innominate vein, vagus, and pleura; on its left, the remains of the thymus gland, and commencement of the left common carotid; and behind it is the trachea. It occasionally divides higher in the neck than at the sterno-clavicular articulation, and may be seen pulsating. Its position must be carefully made out in performing tracheotomy, particularly in children, where the space in which the operation is feasible is very limited, and the structures so close together,

besides which the trachea is very small and movable.
It sometimes gives off a branch to the thyroid body
(the thyroidea ima), which might lie immediately over
the site of the deep incision in tracheotomy; and occasionally it gives off thymic and bronchial branches.

Ligature of the Innominata.—In order to expose the
vessel for the purpose of ligaturing it, it must be drawn
out as much as possible from the deep position it occupies,
by raising the shoulders and throwing back the head.
Next, the sternal origin of the sterno-cleido-mastoid is
to be exposed, by an incision along its anterior border,
met by one along the upper edge of the sternum; this
tendon is to be divided, and the underlying origins of
the sterno-hyoid and thyroid muscles carefully divided.
Beneath these muscular fibres is a plexus of veins, chiefly
derived from the inferior thyroid, which must be hooked
aside. The deep cervical fascia is next to be cautiously
scraped through, and the origin of the common carotid
exposed, which vessel serves as the guide to the trunk
of the innominata. The left innominate vein must be
drawn down, whilst the right innominate and internal
jugular veins, with the vagus nerve, are to be drawn
outwards. The needle must be passed from *below, upwards* and *inwards,* taking care that it is kept close to
the vessel, to avoid wounding the right pleura, or trachea
which lie behind it. The ligature should be applied as
high up as possible.

*Collateral Circulation developed after Ligature of the
Innominata.*—The right side of the head and neck would
be supplied with blood by the inosculation of the carotids
with those of the opposite side; and the circulation in
the right subclavian would be re-established, by means
of its intercostal branch inosculating with the first aortic

intercostal, assisted by the internal mammary and intercostal arteries, inosculating with the long thoracic, superior thoracic, and acromio-thoracic, and by the inosculation between the superior thoracic and deep epigastric.

SURGICAL RELATIONS OF THE ŒSOPHAGUS IN THE NECK.

The œsophagus is occasionally the seat of operation,
such as for the removal of some foreign body, of the passing of bougies in cases of stricture, or introducing the
tube of a stomach-pump. The operation of *œsophagotomy*
is required but rarely, but it may be had recourse to in
such cases as impaction of foreign bodies, when the substance can neither be pulled out through the mouth, nor
pushed downwards into the stomach, or in such a case as
where a substance might be removed through a longitudinal wound in the œsophagus, but could not be drawn
through the more constricted portion of the tube—as,
for instance, a set of false teeth, &c. In the neck, it
commences as a constriction below the pharynx, having
at this point the cricoid cartilage in front, and the fifth
cervical vertebra behind it.

Just at first, it lies in the mesial line of the body, but
as it approaches the root of the neck it inclines towards
the left side.

Relations.—In front of it lies the trachea, and after it
tends to the left side, the left lobe of thyroid body, and
the thoracic duct.

Behind, the cervical vertebræ and left longus colli
muscle.

At the sides, the common carotid vessels, particularly
the left, the thyroid body, and the recurrent laryngeal
nerves.

Great care must be taken, in passing bougies or tubes through the mouth into the œsophagus, to keep the end of the instrument well against the spine, and to use very gentle pressure, as false passages are readily made (especially where there has been any disease) into the pleural cavity, posterior mediastinum, or pericardium.

The operation of œsophagotomy is thus performed.   An incision is to be made on the left side of the neck, about four inches long, along the anterior border of the sterno-mastoid, as though for ligaturing the common carotid artery above the crossing of the omo-hyoid.   The omo-hyoid, sterno-hyoid, and sterno-thyroid muscles are to be drawn downwards and inwards, and the sheath of the vessels, uninvolved, drawn outwards ; the œsophagus is then seen at the bottom of the wound, when a longitudinal incision is to be made upon the foreign body or bougie as it lies in the tube.

The *structures to be avoided* are—the sheath of the vessels, the thyroid vessels, the thyroid body, and the laryngeal nerves.

The occipital portion of the side of the neck, that above the crossing of the omo-hyoid, possesses few points of surgical importance beyond it being the seat of tumors. Its boundaries are—in front, the sterno-mastoid ; behind, the trapezius ; and below, the omo-hyoid ; its floor is formed by the upper portion of the anterior scalene muscle, the middle and posterior scalene, the levator anguli scapulæ and splenius colli muscles.   The spinal accessory nerve emerges from the junction of the upper and middle third of the posterior border of the sterno-mastoid and crosses the region obliquely, to enter the trapezius, accompanied by descending muscular branches of the cervical plexus ; the superficial branches of the

9

cervical plexus are seen, passing upwards (small occipital) along the posterior border of the sterno-mastoid,
forwards (great auricular and transverse cervical) across
it, and downwards (descending cervical). There are a
great many lymphatic ganglia along the posterior border
of the muscle, and the integument is very tough and
fibrous.

REGION OF THE NAPE OF THE NECK (POSTERIOR CERVICAL REGION).

This region extends from the occipital tuberosity and
superior curved lines above, to the seventh cervical vertebra below, and is bounded laterally by the trapezius.
The spinous processes of the three last can be felt through
the integument, the seventh receiving the name of *vertebra prominens.* The integument is very tough and
strong, containing a great deal of fibrous tissue, and not
very vascular, and is a favorite seat of abscess or car-
buncle; the subcutaneous cellular tissue contains a good
deal of fat, and is united to the ligamentum nuchæ, a
tough fibro-elastic mass extending from the occipital
tubercle to the seventh cervical vertebra, separating the
muscles on either side of the neck. Beneath the aponeuroses lie the trapezius, the aponeurotic attachment of
which superiorly is blended with that of the sterno-mastoid; and separating these muscles from the succeeding
layer is a dense fascia, continuous with the dorsal aponeurosis, beneath which are found the splenius capitis
and the levator anguli scapulæ, the upper portion of the
rhomboidei, and serratus posticus superior; next the
complexus, the trachelo-mastoid, and transversalis colli.
In the fat and cellular tissue beneath this layer are

several vessels of importance in the maintenance of the collateral circulation after ligature of the subclavian or carotid arteries—namely, the occipital, vertebral, profunda, and princeps cervicis. The *occipital* enters the region between the splenius capitis and obliquus superior, and lies between the splenius and the complexus, and is afterwards distributed to the scalp; its descending branch, the *princeps cervicis*, which inosculates with the vertebral and the *profunda cervicis* branch of the superior intercostal, passes between the complexus and semi-spinalis colli muscles. The profunda cervicis enters the region by passing backwards between the transverse process of the seventh cervical vertebra and the first rib. The *nerves* found here are branches of the suboccipital and the posterior branches of the great occipital, and third and fourth cervical nerves. Beneath the integument are a number of lymphatic ganglia, which become engorged in constitutional syphilis. Beneath the cellulo-fatty layer are the muscles forming the suboccipital triangle, the recti capitis postici, internally, the obliquus superior and the obliquus inferior, the floor of which triangle contains the curve of the vertebral artery perforating the posterior occipito-atlantoid ligament, before it takes its course through the foramen magnum; and between this vessel, and the groove in the atlas, is the trunk of the suboccipital nerve.

Owing to the curvature of the cervical spine, the prominence of the second and third cervical vertebra is seen in the pharyngeal cavity, and can be readily recognized from the mouth.

The cervical spine is liable to dislocation from the great mobility of the articulations; and from the fact that the articular surfaces are nearly horizontal, dislocation without fracture may take place. In the space between the first cervical vertebra and the occiput the cord may be readily reached by a mere puncture. An injury to the cord above the third would implicate the phrenic nerve and paralyze the diaphragm, and give rise to sudden asphyxia, causing instantaneous death.

If the lower portion of the cervical cord be divided there will be paralysis both of the upper and lower extremities, and the respiration would be entirely carried on by the diaphragm. If the injury be opposite the sixth cervical vertebra, there would be only partial paralysis of the upper extremity, owing to that portion of the brachial plexus given off from the cord above this spot being unimplicated.

The cervical spine is occasionally the seat of *spina bifida*, an arrest of development in which the spinous processes or the laminæ are absent or separated, allowing of the bulging of the meninges beneath the integuments. In spina bifida in this region, the spinal cord and its nerves are generally adherent to the parietes of the tumor.

The relation of the *structures passing through the superior aperture of the thorax* would seem to form a suitable connection between the regions of the thorax and neck. Supposing a section be made—a decapitation, in fact—following a plane passing through the upper part of the first dorsal vertebra behind, the upper borders of the first ribs laterally, and the manubrium sterni anteriorly, the following structures would be seen passing from before backwards, between the apices of the lungs. In the middle line are the origins of the sterno-hyoid and sterno-thyroid muscles, and a lax cellular tissue, in which are the remains of the thymus gland and inferior thyroid

veins, the trachea, œsophagus ; and, in the groove separating them, the recurrent laryngeal nerves ; and on the
left side the thoracic duct.   At the sides the internal
mammary vessels, the innominate veins, and on the right
the innominate artery, with the vagus nerve lying be-

FIG. 17.

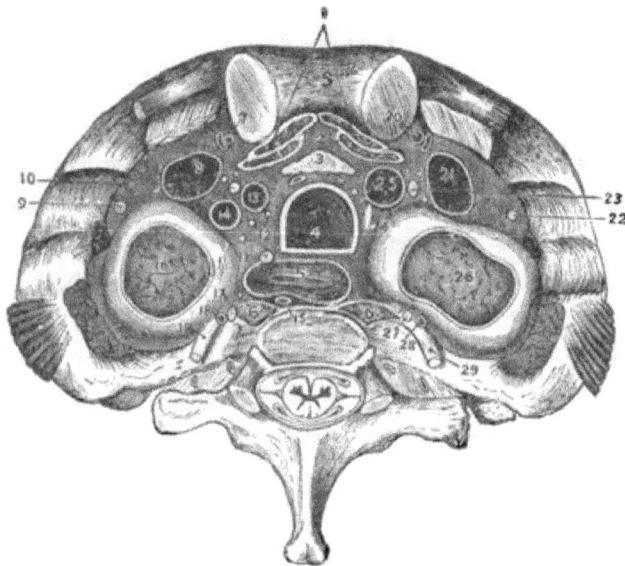

Median Line.—1. Sterno-hyoid muscles. 2. Sterno-thyroid muscles. 3. Remains of thymus gland. 4. Trachea. 5. Œsophagus. 6. Longi colli muscles.
Left Side.—7. Internal mammary artery. 8. Innominate vein. 9. Phrenic nerve.
10. Vagus nerve. 11. Recurrent laryngeal nerve. 12. Cardiac nerves. 13. Left
carotid artery. 14. Left subclavian artery. 15. Thoracic duct. 16. Apex of lung
and pleura. 17. Sympathetic. 18. Superior intercostal artery. 19. First dorsal
nerve. *Right Side.*—20. Internal mammary artery. 21. Innominate vein. 22.
Phrenic nerve. 23. Vagus nerve. 24 Cardiac nerves. 25. Innominate artery.
26. Apex of lung and pleura. 27. Sympathetic. 28. Superior intercostal artery.
29. First dorsal nerve. (HEATH.)

tween it and the innominate vein, and on the left the
common carotid and subclavian arteries, with the vagus
between them, the phrenic and cardiac nerves, the trunk
of the sympathetic, the longi colli muscles, and the superior intercostal arteries and first dorsal nerves.

CHAPTER III.

SURGICAL ANATOMY OF THE THORAX.

IN the region of the thorax it is intended to include that portion of the body comprising the parietes of the chest, which contains the heart, lungs, and the contents of the posterior mediastinum, and is bounded, *superiorly*, by the superior aperture of the thorax (that is to say, the bony ring formed behind, by the body of the first dorsal vertebra, on either side, by the first ribs, and anteriorly, by the upper part of the manubrium sterni); *inferiorly*, by the diaphragm ; and *laterally*, by the ribs and inter-costal muscles, above the limit of the diaphragm, to the exclusion of such structures as are regarded as belonging to the upper extremity—namely, those entering into the formation of the axilla, which region will be hereafter described.

The applied anatomy of the thorax is rather the province of the physician than the surgeon, yet at the same time there is so much in common that, with regard to physical examination, it will be advisable to devote some little space to the subject.

Before going into any details of its structure it is necessary to point out the relations of its contents with reference to the walls of the chest—such in fact as relate to the auscultation or percussion of the lungs, heart, and great vessels. As, however, these matters are to be

found in works specially devoted to the subject, it is
proposed merely to point out the chief anatomical bearings of the contained viscera.

The Lungs.—Presuming the body to be normal, their
position with regard to the thoracic walls is as follows :
The *apices* lie beneath the scalenus anticus muscle and
the subclavian artery, separated by the œsophagus, trachea, and anterior portion of the bodies of the first and
second dorsal vertebræ. The *bases* of each are separated
from the abdominal viscera by the diaphragm, that of
the right being considerably hollowed out by the bulging upwards of the liver, as far up in the thorax as the
fifth rib ; that of the left is hollowed out to a less degree
by the projection of the stomach, spleen, and left lobe of
liver.

The inner margin of the right lung passes vertically
down the middle of the sternum, with a slight inclination to the same side, as far as the sterno-xiphoid articulation.

The inner margin of the left lung lies parallel with that
of the right, about as far as to the fourth costal cartilage, where it passes outwards along this cartilage for a
short distance, and then descends obliquely downwards
and backwards, a little internal to the nipple, nearly as
far as the seventh rib.

The Heart.—The heart lies obliquely, and during expiration is nearly horizontal, its base being to the right
and apex to the left side. The base corresponds to the
interval between the fifth and eighth dorsal vertebræ,
and its apex to a little below the left fifth rib, to the
left of its junction with its cartilage, while the *impulse*
is to be felt in the interspace between the cartilages of
the fifth and sixth ribs, internal to the nipple. Its

upper border corresponds to a line on a level with the upper borders of the third costal cartilages, and its lower border to a line extending between the articulation of the ensiform with the costal cartilage on the right side, to the position of the apex. Hence, for auscultation of the base, the spot is the upper border of the third costal cartilage, and for the apex, a point about two inches below

FIG. 18.

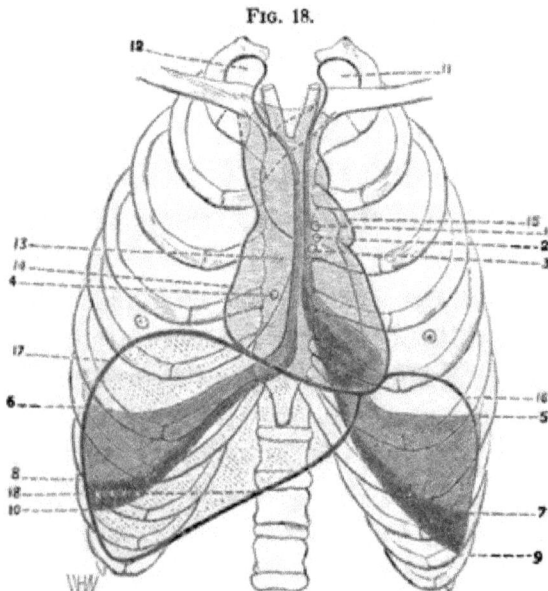

Diagram of the relations of the thoracic viscera to the walls of the chest (altered from ANGER). 1. Situation of pulmonary orifice. 2. Left auriculo-ventricular orifice. 3. Orifice of aorta. 4. Right auriculo-ventricular orifice. 5. Limit of the anterior and inferior border of left lung in complete expiration. 6. Ditto of right lung. 7. Limit of left lung in inspiration. 8. Ditto of right lung in inspiration. 9. Limit of pleura. 10. Ditto. 11. Superior cul-de-sac of left lung. 12. Ditto of right lung. 13. Right auricle. 14. Right auricular appendage. 15. Left auricle. 16. Limit of diaphragm in complete expiration. 17. Ditto, ditto. 18. Ditto, ditto, in complete inspiration.

the nipple, and one inch towards the middle line of the body.

The *right auriculo-ventricular* opening is behind the

centre of the sternum, on a line with the lower margin of the articulation of the cartilage of the fourth rib with it. The *left auriculo-ventricular* opening is about three-quarters of an inch lower than the pulmonary orifice. The orifice of the pulmonary artery is on a line with the space between the junction of the second and the third costal cartilages with the sternum, being to the left and close to that bone.

The *orifice of the aorta* is at the commencement of the ascending portion of the arch, and is on a line with the junction of the third costal cartilage of the left side with the sternum. The *arch* attains the level of the upper border of the second costal cartilage of the right side at its junction with the sternum. (*Note.*—These relations vary slightly in the works of some authors, but those mentioned above will be found correct for practical purposes.)

The region of the walls of the thorax may be conveniently described surgically, as *sternal, costal, diaphragmatic,* and *spinal.*

The *sternal* region consists of the sternum itself and of the structures which immediately cover it—viz., integument, subcutaneous cellular tissue, aponeurosis of pectoralis major, sterno-mastoid, rectus abdominis, and occasionally fibres of the rectus sternalis muscles ; it is surgically of importance, as being the seat of fractures, of dislocations of the clavicle from it, of necrosis from various causes and of growth. The mechanism of the sterno-clavicular articulation, which possesses an inter-articular fibro-cartilage and two synovial membranes, allows of motion in almost all directions, and largely assists in the free play of the shoulder. The peculiarity in the construction and the curved form of the clavicle,

serve to break the effect of shocks or blows upon the
shoulder or upper limb. *On the right side* the sterno-
clavicular articulation is immediately in front of the innominate artery and subclavian vein, whilst on the *left*
it is in relation with the left subclavian vein and the
interspace between the left common carotid and subclavian arteries; the interclavicular space lies immediately
in front of the trachea. Posteriorly and below, the
sternum is covered by the triangularis sterni muscles,
and laterally is in relation with the internal mammary
vessels. The sternum is liable to fractures, and dislocations from it of the clavicle—an accident which may
occur in any direction but downwards, owing to its close
relation with the cartilage of the first rib.

The costal region, which is bounded anteriorly by the
sternum, laterally by the sides of the bodies of the vertebræ, and inferiorly by the diaphragm, presents several
points of surgical importance, since it contains the mammary gland and the intercostal spaces. It is covered in
anteriorly and posteriorly above by structures belonging
to the region of the upper extremity, in front by the
greater and lesser pectoral muscles with their aponeuroses,
and behind by the scapula and the muscles attached to
it, and between the external borders of the scapular and
pectoral muscles, the ribs and the humerus, is a special
region,—the axilla.

The *intercostal spaces* vary considerably in extent:
thus they are larger during inspiration; the most remarkable variation in the interspace is at its middle, as
during expiration the ribs occasionally approximate so
closely that their edges are in absolute contact. The
operation of *paracentesis thoracis* or tapping the thorax,
in pleurisy, emphysema, or the pointing of an hepatic

abscess, is usually performed between the fifth and sixth ribs, just behind their middle, and *never* behind the angle, because of the size of the intercostal arteries at this point, and moreover on account of the thickness of the layers of muscle.

Some prefer a point situated an inch or more below the angle of the scapula, between the seventh and eighth, or eighth and ninth ribs, the instrument being passed just above the *upper* border of the rib, so that the intercostal vessels may be avoided. On the right side, it is advised to puncture through an interspace higher, on account of the position of the liver and diaphragm on that side.

The internal mammary artery belongs properly to the cavity of the chest, lying on the left side in the anterior mediastinum, whilst on the right it is so overlapped by the lung as to be excluded from the space. It is best exposed from the surface, by raising the cartilages of the ribs and the sternum, and is then seen lying upon the pleura at a short distance from the margin of the sternum. It is given off from the subclavian opposite the vertebral, and passes into the thorax in relation with the phrenic nerve, which crosses it anteriorly and then descends internal to it. Two veins, which usually unite to form a common trunk, accompany it. In the upper part of the chest it is covered in by the costal cartilages and internal intercostals, whilst below it lies between the triangularis sterni and the pleura. Its chief inosculating branches are given off at the interval between the sixth and seventh cartilages, and are *musculo-phrenic* and *superior epigastric*.

Wounds of this vessel, especially if situated where it has any bulk, *i. e.*, between the first and seventh rib, are serious. It can be *tied* easily in the *first three* intercostal spaces, by making an oblique incision about two inches long from without inwards, and at about three or four lines from the border of the sternum, and the *structures divided in reaching it* are—integument, cellular tissue, origin of pectoralis major, internal intercostal muscle. Owing to its free anastomoses, of course both ends of the wounded vessel require ligature. The close neighborhood of the ribs to the lungs and pleura, liver and diaphragm, render fractures by direct violence often very serious from puncture of these structures by the fragments. In indirect violence the rib is broken, as a rule, near the angle, and there, save danger from punctures, wounds of the intercostal arteries are rare; these vessels are difficult to secure, from their position.

Each intercostal artery is accompanied by a vein and nerve, the nerve being superior to the artery in the upper intercostal spaces, but below after the fourth or sixth space.

They are protected from pressure whilst the intercostals are acting, by being inclosed in tendinous bands, which are attached to the ribs.

Mamma.—The breast is situated in front of the pectoralis major, towards the lateral aspect of the region of the chest, and corresponds to the interval between the third and seventh ribs (the *male* nipple lies on the fourth rib). It consists of gland and fibrous tissues arranged in lobes; these are very numerous, the septa between them being filled up by fat. Each lobe is again divisible into lobules, which are connected by areolar tissue, bloodvessels, and ducts. The lobules open into the lactiferous ducts, which uniting form larger ones, terminating in an excretory duct, and are generally from fif-

teen to twenty in number; these converge towards the
areola, and beneath the nipple become dilated into sinuses, before perforating its summit by separate orifices.
The mammary gland is separated from the pectoralis
major and minor, the serratus magnus, and its sheath by
a layer of cellular tissue which allows of the free movement of the gland over it,—an important point in the
diagnosis of breast tumors.

The breast is freely supplied with vessels, the arteries
being derived from the internal mammary, the long
thoracic from the axillary, and the intercostals; the
veins are both superficial and deep: the superficial are
seen beneath the subcutaneous cellular tissue, and are
very much distended during pregnancy, whilst the deeper
ones follow the course generally of the arteries.   An
anastomotic circle of veins is seen around the base of
the nipple.   The lymphatics are also arranged as superficial and deep: the former are immediately beneath the
integument and pass into the axillary glands; the latter
set accompany the galactiferous tubes, and pass into the
cellular tissue beneath the gland, also to join the glands
in the axilla and the intrathoracic ganglia.   The nerves
are derived from the brachial and cervical plexuses and
from the intercostals.

The breast is the seat of many forms of tumors which
necessitate its removal, and its great vascularity gives
rise to severe hemorrhage during such operations; it is
advisable, therefore, in such cases to make the *inferior*
incision first, to avoid, if possible, any complication,
owing to the parts being obscured by blood.   Mammary
abscesses ought to be opened *vertically* (to avoid "pocket-
ing" and the formation of sinuses), freely and deeply,
to insure the exit of all matter.   Moreover, these in-

cisions should, if possible, be made parallel to the course of the galactiferous ducts, if near the nipple, in order to avoid cutting across them. Guided by the fascial envelope, collections of matter connected with the mammary gland will occasionally point, and require opening at the anterior border of the axilla.

The diaphragmatic region, or floor of thorax, is formed by the diaphragm, which constitutes the septum between the thoracic and abdominal viscera, and is the muscle of normal respiration. Its height, or the amount of encroachment upon the thorax during ordinary respiration, depends in some measure upon the amount of distension and the size of those abdominal viscera in immediate relation with it—viz., the stomach, intestines, and the liver. During *normal expiration* the right arch ascends to the level of the *fifth* rib. *Forced expiration* brings the right arch of the muscle—that above the liver—to a level with the fourth costal cartilage in front, with the fifth, sixth, and seventh ribs at the side, and with the eighth rib behind. The left arch is lower than the right by two ribs. During *forced inspiration* the muscle descends to the level of a line extending from the ensiform cartilage to the tenth rib.

The under surface of the diaphragm is perforated by three large foramina: (1) The aortic, situate between the pillars of the muscle and spinal column, transmits the aorta, the thoracic duct, and the vena azygos major. (2) The caval, quadrilateral in shape and incapable of constriction, transmits the inferior vena cava. (3) The oesophageal, elliptiform in shape and capable of constriction, transmits the oesophagus, also the vagi nerves,

the left being in front. On either side of the attachment to the xiphoid cartilage is a space where the muscular tissue is wanting, so that between the abdominal
cavity and that of the anterior mediastinum there is a
communication, filled in by a little cellular tissue,
through which pass some lymphatics from the liver,
and occasionally diaphragmatic hernia. Collections of
pus forming in the thoracic cavity may find their way
through these spaces; moreover, as the diaphragm separates the right lung from the liver, abscesses forming in
this latter viscus may either be discharged by the bronchi or into the thoracic cavity.

Penetrating wounds of the diaphragm are serious,
partly on account of hemorrhage, and from the fact of
viscera, both of the thoracic and abdominal cavities,
being implicated; and partly on account of the intimate
connection with it of large serous cavities, the pleuræ,
the peritoneum, and the pericardium.

Occasionally *paralysis* of the diaphragm occurs with
ascites, or may be owing to a wound in the neck or spinal column, implicating the phrenic nerve.

An approximation of the course taken by a bullet or
a weapon penetrating the walls of the thorax from its
anterior or lateral aspects, may be gathered from the
following facts: That the *heart* would be reached by a
wound traversing the chest at right angles, above the
sixth rib, and that its apex lies about an inch and a
half from the surface. Wounds in the mesial line
would involve the heart and great vessels, whilst more
laterally they would implicate the lungs; the position
of the trunk and branches of the internal mammary
render penetrating wounds of the inferior intercostal
spaces near the sternum very serious.

A wound penetrating the sternum, on a line with the nipple, and striking upon the vertebra at right angles with the axis of the body, would traverse three cavities of the heart, two ventricles, and the left auricle.

If an instrument traversed the lower intercostal spaces during inspiration, it would wound the base of the lung, the diaphragm, and corresponding abdominal viscera, but during expiration would avoid the lung.

Spinal Region of Thorax.—The posterior wall of the thorax is formed by the dorsal spine and its coverings, and is represented by a region bounded superiorly by the first, and inferiorly by the last dorsal vertebra, and laterally by the angles of the ribs.

The integument is very thick and dense, and contains a quantity of sebaceous follicles; the subcutaneous cellular tissue is intimately united to the spinous processes of the dorsal vertebra (which can be plainly felt throughout the region), thus preventing fluid infiltrating or pus appearing superficially, except along the sides of these processes. The dorsal aponeurosis invests the anterior and posterior surfaces of the trapezius and latissimus dorsi, uniting superiorly with the cervical, and inferiorly with the lumbar aponeurosis, splitting to inclose the serrati postici; it is attached to the spinous processes of the dorsal vertebræ and tubercles of the ribs, inclosing all the muscles of the back, and separating the superficial from the deep layer. The first layer of muscles consists of the lower part of the trapezius and the latissimus dorsi, the former overlapping the latter. Beneath these muscles lie the rhomboidei and serrati postici. The deep layer consists of the sacro-lumbalis internally with its necessary muscles,

and the longissimus dorsi externally also with its accessory muscles; between them superiorly, the lower portions of the splenius colli, complexus, and transversalis
colli; deepest of all, the transverso-spinales and levatores
costarum.

Abscesses connected with caries of the transverse processes, or laminæ of the dorsal vertebræ, gravitate into
the inferior portion of the region, and often extend to
the sides of the ribs or into the axilla, instead of becoming superficial, on account of the strong fascial lamina
between the superficial and deep muscles.

A small triangular space uncovered by muscle exists
at the point where the trapezius and latissimus dorsi
diverge. The space is bounded by the two muscles just
named towards the spine, whilst externally it is completed by the inferior angle of the scapula. From the
fact of its being uncovered by muscle it is available for
auscultation.

The *vessels* met with are dorsal branches from the intercostals and the posterior scapular, and the *nerves* are
the dorsal branches of the spinal, with some few branches
from the cervical plexus and spinal accessory.

With regard to the skeleton of this region, the arrangement of the dorsal vertebræ renders their dislocation extremely difficult, and when it does occur it would
seem to be invariably associated with fracture of some
portion of the spinal column; the spinous processes from
their position are often fractured.

The dorsal spine is very often the seat of caries of the
bodies of the vertebræ, or disease of the intervertebral
substances, inducing angular curvature and paralysis of
the lower limbs from pressure on the cord. When the

disease depends upon caries of the bodies of the vertebræ and abscesses form, the course taken by the pus will depend upon the part of the spine which is the seat of disease ; and it generally escapes beneath the pillars of the diaphragm, and passes beneath the fascia, along the side of the aorta and iliac arteries, pointing in the abdominal parietes *above* Poupart's ligament. If the abscess depends upon disease of the lower dorsal vertebræ, the pus is directed forwards by the sheath of the psoas muscle, and points *below* Poupart's ligament, in the front of the thigh, and external to the vessels. The pus sometimes passes backwards, forming dorsal abscess. If it gets into the subperitoneal areolar tissue in the pelvis, it may find its way into the perineum by the side of the rectum, or pass out of the great sciatic notch, and appear in the region of the great trochanter. From the continuity and density of the fascia of the leg, such collections of matter may make their way into the popliteal space, or even along the side of the tendo-Achillis.

Injuries to the cord in the dorsal region, if below the second dorsal vertebra, do not affect the upper extremity, but the respiration is greatly affected, owing to implication of the nerves supplying the intercostals and abdominal muscles ; moreover, there is paralysis of all the parts supplied by the nerves below the seat of injury.

Region of the Cavity of the Thorax.

As this region can scarcely be considered as within the province of *surgical* anatomy, it is proposed therefore to allude to it as shortly as possible. Its practical bearings to the surgeon seem rather to be upon the relations of the contents to the parietes, which have been already discussed, both as far as physical diagnosis and injury are concerned ; and considerations with respect to aneurism of the aorta or its great trunks are subjects rather for systematic surgery. Since operative proceedings on the œsophagus, trachea, or great vessels are instituted in the neck, the chief anatomical bearings of these structures will be found described in the chapter on the Surgical Anatomy of the Neck.

The cavity of the thorax is most conveniently described for reference as divisible into two pleural cavities, separated in the lower four-fifths of the chest, by the three mediastina, and in the upper fifth by the great vessels springing from the arch of the aorta, a region called by Professor Wood, the *cervico-thoracic*. The posterior wall of this division, he makes the three upper dorsal vertebræ and their intervertebral substances, and the intervertebral substance between the third and fourth the anterior—the manubrium sterni, and upper fourth or fifth of the anterior mediastinum ; on either side the apices of the lungs and pleuræ.

The *anterior mediastinum* is bounded as follows : Anteriorly, by the sternum and left costal cartilages ; posteriorly, by the pericardium, for the lower three-fourths or four-fifths, and for the rest, by the cervico-thoracic region ; laterally, by the pleura. The *middle mediastinum* contains the heart and ascending portion of the aorta and arch, phrenic nerves and vessels, and is limited behind by fibrous pericardium and the obliterated ductus arteriosus.[1] The *posterior mediastinum* is limited above

---

[1] *Vide* Professor Wood, F.R.S , on Relations of Aorta : " Journal of Anatomy and Physiology," vol. iii.

by the left portion of the arch, **below** by the lesser muscle of the diaphragm ; laterally, pleural cavities, root of **lungs,** and ligamenta lata pulmonum.

It contains the bifurcation of the trachea and bronchi, vagi, œsophagus, hinder part of root of lung and recurrent laryngeal nerve of left side, descending aorta, azygos major vein, thoracic duct, and great splanchnic nerves.

.

CHAPTER IV.

SURGICAL ANATOMY OF THE UPPER EXTREMITY.

Region of the Shoulder.

Surface Markings.—The convexity of the shoulder is due to the deltoid muscle, and the globular head of the humerus. The bony processes, the coracoid, acromion, spine of scapula, globular head of humerus, and the entire extent of the clavicle, can be readily felt, and their exact relation with regard to each other should be noticed and compared, their respective bearings with the other bony prominences of the upper extremity carefully studied both at rest and in action. The precise relations of these *surface* marks is of the utmost importance in the diagnosis of fracture, dislocation, or other injury to the shoulder-joint, and moreover they serve as guides for the direction of the knife in amputations or excisions. When the arm hangs along the side with the palm turned forwards, the acromion, epicondyle, and styloid process of radius externally, and internally, the head of humerus, epitrochlea, and styloid process of ulna, correspond exactly, and their mutual relations are to be noted in every position of the joint.

Anterior and Lateral Aspects of the Shoulder : Surface Markings.—*Anteriorly,* immediately below the clavicle, is a fossa in which the pulsations of the first part of the axillary artery can be felt, and which hollow may be obliterated by the presence of axillary tumors or dislocation of the humerus forwards.

Externally, the roundness of the shoulder is formed by the deltoid muscle, beneath which, and below the overhanging process of the acromion, can be felt the globular head of the humerus. This portion of the bone is sometimes very large in proportion to the articulation, and might be mistaken for a displacement. In any dislocation, however, the rotundity of the shoulder gives place to a characteristic flattening; besides impairment of the movements of the joint. Along the anterior border of the deltoid is a groove between it and the upper fibres of the pectoralis major, in which lie the cephalic vein and a branch of the acromio-thoracic artery. The skin of this region is very thick, and glides easily over the underlying tissues, owing to the presence of bursæ, which are in some individuals more or less developed, according to the use made of the shoulder by their occupation or work—thus, by carrying a ladder or a hod of mortar, &c.

To the upper border of the clavicle are seen attached, along its sternal portion, the outer fibres of the sterno-cleido-mastoid, and along its acromial portion the trapezius.

Below are the attachments of the pectoralis major internally, and externally the deltoid. The clavicle can be felt in its whole extent, owing to its subcutaneous position, and from being so exposed is liable to fracture. Again, its structure, its curves, which intensify the shocks of indirect violence, and the want of support posteriorly, all contribute towards the frequency of the accident.

If this fracture takes place at its centre, the displacement of the outer fragment downwards, forwards, and
inwards, is due to the weight of the arm, and if put in
action, to the deltoid, pectoralis major, and subclavius.
The inner fragment is occasionally tilted upwards, but
only when the clavicular fibres of the sterno-cleido-
mastoid are in action, as in rotation of the neck, the
strong ligamentous attachment to the rib (costo-clavic-
ular), and perhaps the clavicular fibres of the pectoralis
major, retaining it in place.   The slightness of the displacement, on fracture of the acromial and sternal extremities, is also due to their strong ligamentous attachment.   The clavicle is separated from the upper part of
the axilla and its contents by the subclavius muscle
and its aponeurosis; behind its sternal extremity is the
junction of the internal jugular vein and subclavian vein
running closely along it; behind its upper border are
the supra-scapular vessels, which are liable to be injured
in fracture of the bone or operations upon it.

External Aspect of the Shoulder.—The deltoid muscle
forms the external boundary of the region of the shoulder.
The subcutaneous tissue contains the terminal twigs of
the acromial branches of the descending clavicular
nerves, and of the cutaneous branches of the circumflex
nerve and artery.   The aponeurosis of the deltoid completely covers the muscle, and is continuous with that of
the upper arm and axilla, sending down septa between
the bundles of the muscular fibres, and passing beneath
it, is continuous with the deep fascia.   The deltoid itself,
arising from the outer half of the clavicle and the lower
border of the spine of the scapula, after the convergence
and interlacement of its fibres, is inserted into the rough-
ened surface on the outer aspect of the humerus.

Parts beneath the Deltoid.—Immediately beneath this muscle is a large bursa, often multilocular, which lies between it and the convex head of the humerus and the acromion process; the insertions of the supra- and infra-spinatus and teres minor muscles into the greater tu-

FIG. 19.

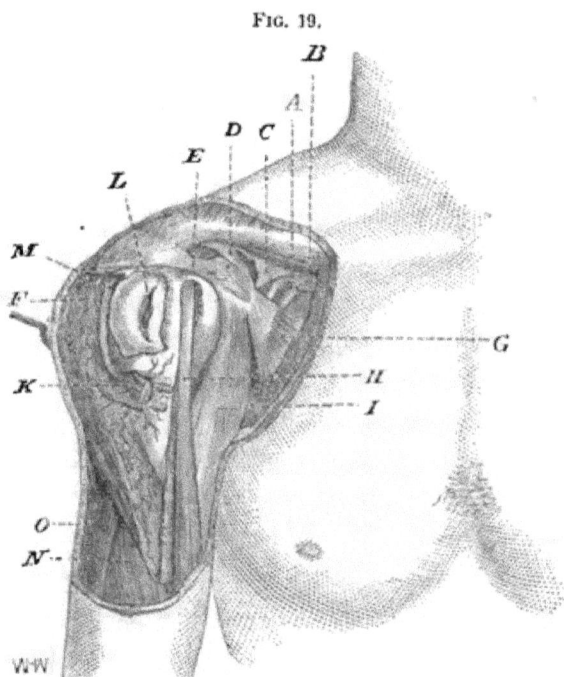

A. Axillary artery, B. Axillary vein. C. Costo-coracoid membrane. D. Coracoid process. E. Coraco-acromial ligament. F. Deltoid cut and pulled back. G. Pectoralis minor. H. Long head of biceps. I. Short head of biceps and coraco-brachialis. K. Circumflex vessels and nerve. L. Head of humerus seen through capsule, upon which is a portion of the bursa between it and the deltoid. M. Posterior portion of this bursa between the deltoid and scapular muscles. N. Brachialis anticus. O. Triceps. (Altered from ANGER.)

berosity, and passing round the surgical neck, the posterior circumflex vessels and circumflex nerve; the bicipital groove, in which lies the long head of the biceps and the anterior circumflex artery; the coracoid process, with the short head of the biceps and coraco-brachialis attached to it, a quantity of loose cellular tissue, and the capsular ligament. Severe blows upon the shoulder frequently cause effusion into the bursa above named, and render diagnosis of extreme difficulty, and suppuration in its cavity may be mistaken for disease of the articulation. In the case of effusion of blood into the cavity, the posterior border of the axilla will become discolored a few days after the injury.

Posterior Aspect of the Shoulder : Surface Markings.— The spine of the scapula is seen directly beneath the skin, as a furrow in muscular persons, and as a prominent ridge in emaciated subjects, terminating in its broad acromion process, which overhangs the articulation. Above and below the spine are the supra- and infra-spinous fossæ, filled in by muscle, the superior fuller than the inferior, owing to the attachment of the trapezius along the upper border of the spine. The inferior angle of the scapula is just beneath the surface, having sometimes fibres of the latissimus dorsi attached to it, which muscle, as it sweeps forward towards the arm, forms the posterior border of the axilla ; occasionally the scapula becomes dislocated over the upper edge of this muscle, or rather the muscle slips beneath the scapula.

The supra-spinous fossæ contains the supra-spinatus muscle, between which and the bone lie the supra-scapular vessels and nerve, the nerve passing below the supra-scapular ligament and the vessels over it ; these vessels, after supplying the supra-spinatus muscle and the nerve, wind round the root of the acromion into the infra-spinous fossa, and there the important inosculation takes place between the dorsalis scapulæ and posterior

11

scapular, which becomes greatly developed after ligature of the subclavian in the third part of its course, playing a chief part in the maintenance of the collateral circulation for the supply of the arm. In front of the scapula, between it and the ribs, is the subscapularis, a multipenniform muscle, the full development of which gives such an appearance of depth to the thorax in muscular persons. Beneath the tendons of the muscles inserted into the greater and lesser tuberosities, are large bursæ, occasionally communicating with that beneath the deltoid, and with the synovial membrane of the shoulder-joint; inflammation or suppuration of which may be mistaken for glandular inflammation or for axillary aneurism.

Articulation of the Shoulder-joint.—The ellipsoidal articular extremity of the humerus is very large in comparison with the glenoid cavity, which is ovoid in shape, larger below than above; and it is surgically of importance to note that, not only are the articular surfaces of the humerus and scapula here in contact, but that the head of the bone is in immediate relation with the arch formed by the coracoid and acromion processes and the coraco-acromial ligaments—an approximation which is due to the action of the deltoid, in *atrophy* of which muscle there is a considerable interspace between these points. The capsule of the joint is materially strengthened by fibrous expansion from the tendons of those muscles which are in *immediate* contact with the articulation—viz., the supra- and infra-spinati, teres minor, subscapularis, and long head of biceps and triceps, and by bands passing from the coracoid process. The synovial membrane lining the capsule is prolonged beneath the subscapularis muscle, and into the bicipital groove.

The movements of which the humerus is capable in the glenoid cavity are very varied, and, with a view to studying the action of the muscles in dislocation and fracture of this bone, and in their diagnosis, may be classified as follows: The humerus is *raised* by the deltoid, supra-spinatus, long head of biceps, and coraco-brachialis; *depressed* by the pectoralis major, latissimus dorsi, teres major, and subscapularis; *brought forward* by the pectoralis major, anterior fibres of deltoid, coraco-

FIG. 20.

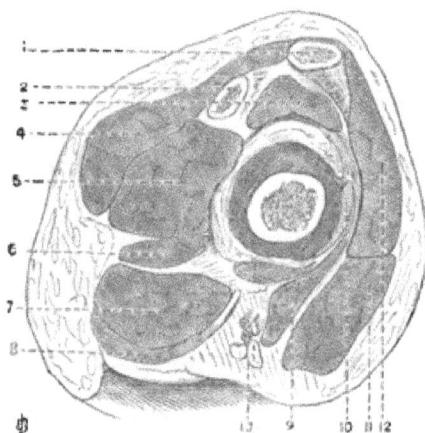

Diagrammatic section through right shoulder-joint, showing structures in contact with it. 1. Clavicle. 2. Acromion. 3. Supra-spinatus. 4. Trapezius. 5. Infra-spinatus. 6. Teres minor. 7. Teres major. 8. Latissimus dorsi. 9. Coraco-brachialis and short head of biceps. 10. Tendon of subscapularis blended with the capsular ligament. 11. Pectoralis major. 12. Deltoid. 13. Axillary vessels and nerves.

brachialis, and short head of biceps; *drawn backwards* by the latissimus dorsi, teres major, long head of triceps, posterior fibres of deltoid, supra-spinatus, and teres minor; *rotated inwards* by subscapularis, teres major, latissimus dorsi, pectoralis major, and anterior fibres of deltoid; *rotated outwards* by the infra-spinatus, teres minor, coraco-brachialis, and posterior fibres of deltoid.

The great power of all these muscles upon this joint, which owes its great extent of motion to the shallow glenoid cavity and large head of humerus, and the laxity of its capsule, favors dislocation under certain circumstances ; and were it not for the bony arch formed by the coracoid and acromion processes, and the ligament between them, the long head of the biceps passing through the capsule over the head of the bone and blending with the glenoid ligament, and the mobility of the scapula, such accidents would be still more frequent.

The anatomical position of the head of the humerus with regard to the neighboring bony structures after its dislocation may be generalized thus,—*subglenoid, subclavicular, subcoracoid, subspinous.*

In the condition termed *subglenoid*, the head of the humerus rests on the inferior border of the scapula, below the glenoid cavity, between the subscapularis and long head of triceps ; and the peculiar numbness in the hand and arm, and frequent coldness and œdema in the limb, are due to pressure upon the brachial plexus and axillary vessels. In the *subclavicular* variety, the head of the bone lies below the clavicle, internal to the coracoid process, upon the second and third ribs, and beneath the pectoral muscles. In the *subcoracoid*, the head of the bone lies deeply in the upper and inner part of the axilla, below the coraco-brachialis and pectoralis muscles. In the *subspinous* form, the head of the bone lies behind the glenoid cavity, below the spine, and between the infra-spinatus and teres minor muscles. These may be regarded as the *complete* forms, which are, of course, liable to modifications.

That portion of the bone belonging to the region of the shoulder—that is to say, as low down as the inser-

tion of the deltoid—is liable to fracture of the anatomical neck, which is *intra-capsular*, and to **fracture of the**
surgical neck, which is *extra-capsular*. In the former
there is little or no displacement due to muscular action;
in the latter case the upper fragment is drawn up slightly
by the supra- and infra-spinatus, teres minor, and subscapularis; the lower fragment is drawn inwards by the
pectoralis major, latissimus dorsi, and teres major,
whilst the deltoid draws it obliquely **from the** side of
the body.

In cases of fracture of the anatomical neck, with separation, it **may necrose,** owing to there being no means
of vascular supply to the fragment, and if not, it may
be inferred that impaction has occurred. **At any rate,**
whatever amount of repair does take place is due **to the**
lower portion of the shaft of the bone. **In impacted**
fracture the axis of the bone is obviously altered, and
there is a slight cavity beneath the acromion, owing to
shortening. The upper articular extremity **unites with**
the shaft at about the *twentieth* year.

The upper epiphysis is sometimes separated in infants,
in consequence **of the carelessness of nurses** in lifting
them suddenly **up by the arm, giving** rise to most serious mischief.

The landmarks already described in the **superficial**
examination of the shoulder are of the greatest importance in the performance of the operations of *amputation
at the shoulder-joint*, and of *excision of the head of the
humerus*. **A** great number of methods are described in
works on surgery for the accomplishment of the disarticulation, but it will suffice to mention two only—viz.,
that of the operation by lateral flaps, and the oval
method, but only as far as anatomy bears **upon the**

surgical processes. In practice, the accurate knowledge
of the anatomy of the parts alone must guide the surgeon, as it is impossible to lay down rules for an amputation to meet the exigencies of every case. For rapidity
of execution, the former may be practiced, and it is based
upon the following anatomical considerations. It is de-
sired to direct the point of the knife in such a manner
that it may most readily and easily transfix the structures
above the joint (and in skilful hands the joint as well),
cut one flap, disarticulate, and expose a second flap, and
so leave all large arterial and venous trunks for division
until the last moment. The point to be felt for from the
surface, which serves as a guide for the point of the
knife to make for, is a spot between the acromion and
coracoid processes and below the coraco-acromial ligament.
Posteriorly, and below, the guide for the entry or emergence of the knife, is the posterior margin of the axilla,
just in front of the tendons of the teres major and latissimus dorsi; it is then pushed upward and forwards
—the elbow being moved outwards and upwards, in
order that the head of the bone may be as low down in
the glenoid cavity as possible—through the structures
close to the bone, until it emerges at the point above indicated, and a large external flap cut; the joint is next
opened, and the attachments of the muscles around it
being divided, the elbow is carried in front of the chest,
and the head of the bone pushed backwards to put the
tendons on the stretch; after the disarticulation is effected, a posterior flap of about the same length is to be
made, in which lie the vessels and nerves.

   *Contents of Flaps after Disarticulation at Shoulder-joint
by Transfixion.*—The external flap thus fashioned will
contain,—integument, the posterior fibres of the deltoid,

circumflex vessels and nerve, tendons of latissimus dorsi, teres major, and teres minor; the internal flap,—the sub-scapularis, long head of biceps, coraco-brachialis, anterior fibres of deltoid, pectoralis major, axillary vessels and nerves, and integuments.

In the *oval* method the articulation is exposed, by making an incision about two inches long down to the bone, immediately below the acromion process, and a curved incision from this point on either side, each inclosing a semilunar flap, to the anterior and posterior folds of the axilla. Disarticulation is next effected.

The most anatomical as well as the best method of *excising* the head of the bone, is by exposing it by a vertical incision extending from a point just external to the tip of the coracoid process, corresponding to the position of the bicipital groove; an incision which has the advantage of avoiding the circumflex vessels and nerve. The long tendon of the biceps should be preserved, if possible, to assist in the movements of the resulting false joint.

SURGICAL ANATOMY OF THE AXILLA.

Surface Markings and External Form.—When the arm lies against the wall of the chest the area of the axillary space becomes confined, and for anatomical considerations ceases to exist; also when extended beyond a right angle, the head of the humerus projects into the space and obliterates its fold. But when the arm is raised to about an angle of 45°, and the muscles contract, the depth of the fold of the axilla is most marked.

The *boundaries* of the axilla, which are seen beneath the skin, are,—*anteriorly*, the lower margin of the pectoralis major muscle, rounded and muscular, but becoming short and tendinous as it approaches the humerus; *posteriorly*, the lower edge of the latissimus dorsi muscle; *internally*, the chest-wall; *externally*, the arm. The axillary artery is readily felt along the external boundary, and may be here compressed against the bone as it lies in the third part of its course, and it will be observed that this vessel follows the course of the arm in whatever position it takes. . The *base* is formed by the integument, which is fully provided with hair-bulbs and sebaceous follicles.

As the axilla would, in most cases, be attacked surgically from below, that is, from its base towards its apex, it will be found to be advisable to describe its relations and contents as they would be met with in this direction.

Dissection.—The arm is to be raised to a right angle with the trunk, and the palm of the hand turned forward. The integument being removed along the boundaries of the base, the subcutaneous cellular tissue is first met with, containing a good deal of reddish fat in its meshes; next, an aponeurosis, which is continuous in front with the sheath of the pectoralis major; behind with that of the latissimus dorsi; externally with the brachial aponeurosis, and internally with that covering the serratus magnus. On removing this aponeurosis the axillary space is opened; a large quantity of loose fat and cellular tissue and a quantity of lymphatic glands are seen filling up the interspace between the thorax and the arm.

Lying in this cellular tissue, and bridging across from the arm to the chest, will be seen a good many nerves, the intercosto-humeral, which, in some subjects, form almost a plexus, supplying the skin of the base of the axilla; together with some branches of the axillary artery, the long thoracic and its veins passing downwards and forwards towards the anterior inferior aspect of the space, besides a considerable number of branches to the glands (alar thoracic).   On removing this cellular tissue the walls of the axillary space can be made out.

The *internal* wall, slightly convex, is formed by the first four ribs and their intercostal muscles, and the first five serrations of the serratus magnus, upon which lie the posterior thoracic nerve, the superior thoracic, and long thoracic branches of the axillary artery, with their corresponding veins.   The *external* wall, formed by the scapulo-humeral region, is the most important, as on it lie the great vessels and nerves in their fascial envelope; and the fact of the close adherence of these structures to this wall of the axilla is of great value to the surgeon in the extirpation of tumors or the opening of abscesses, which fortunately as a rule lie along the inner wall.   On either side of the bicipital groove are inserted the tendons of the pectoralis major and teres major, the latter being internal, and a little anterior and external is the tendon of the latissimus dorsi.   Lying in this groove and inclosed in a prolongation of the synovial membrane of the joint is the long head of the biceps itself; and most internally are seen the conjoined fibres of the coraco-brachialis and short head of biceps, the inner border of the former being the guide to the vessel; the insertion of the tendon of the subscapularis, and origin of the long head of the triceps.   Above the tendon of the teres major the lower portion of the capsule of the joint is visible.

The *anterior* wall is formed by the pectoralis major

12

and minor and their aponeuroses, and the lower border of the former covered by the integuments constitutes its anterior inferior margin.   In the *female* this margin is hidden by the mammary gland, which overhangs it. The position of this gland is not influenced by the movements of the shoulder upon the trunk, as the cellular membrane between it and the anterior layer of the sheath of the pectoralis major permits of the free motion of the muscle beneath it; but in the case of scirrhus, owing to infiltration of the tissues, the movements of the pectoral are made with great pain and difficulty.   On the posterior surface of these muscles are seen the acromio-thoracic vessels and external and internal anterior thoracic nerves.

The *posterior* wall is formed by the teres major and latissimus dorsi muscles below, and by the subscapularis above with their vascular and nervous supply.

The *apex* of the axilla may be referred to the coracoid process, though more correctly to the aperture between the clavicle, upper border of scapula, and first rib, with their muscular coverings.   The cellular tissue of the cavity of the axilla becomes continuous with that of the subclavian region at the apex by enveloping the vessels that pass through this interspace.

The axillary artery, axillary vein, and brachial plexus form a vasculo-nervous cord, bound together by a dense cellular sheath which is placed on the outer wall of the space, and lying along the inner border of the coraco-brachialis muscle; the pulsations of the artery are felt at the apex (if the pectoralis major is relaxed), and at its lower portion, and in thin persons, the *cord* of the median nerve is usually seen stretched over it when the arm is raised from the side.   The *course* of this vessel is indicated by a line passing through the axilla, drawn from about the centre of the clavicle to the inner border of the coraco-brachialis muscle.   It lies in an envelope of nervous cords,—the median, musculo-spiral, musculo-cutaneous, ulnar, and internal cutaneous.   The axillary vein is very large, and lies internal to and a little in front of the artery, and in ligature of the artery it is seen first, and must be drawn to one side.   It is adherent to the cellular tissue and by fascial attachment to the coracoid process, and if wounded is liable to gape considerably, and thus admit air, an accident which has happened in removing axillary tumors.

The branches of the axillary artery are usually given off in the following order : the *thoracica suprema*, which runs along the upper part of the inner wall ; the *acromiothoracic*, sending branches to its anterior wall ; the *long thoracic*, lost on the thorax and mammary gland ; the *subscapular*, which descends obliquely along the inferior border of the subscapularis, and is distributed to the muscles of the posterior wall, one large branch in particular, the *dorsalis scapulæ*, passing to the dorsum scapulæ in the triangular interval between the two teres muscles and long head of triceps ; the *posterior circumflex* passes through the quadrilateral space formed by the two teres muscles, long head of triceps, and humerus, and winds round the neck of the humerus, supplies the deltoid, and is accompanied by the circumflex nerve and veins ; the *anterior circumflex*, a small branch, is supplied to the articulation beneath the coraco-brachialis and biceps, and inosculates with the former.

Thus, if the vessel be normal, the acromio-thoracic, and thoracica suprema are given off above the pectoralis minor, the external mammary about opposite its middle,

and the dorsalis scapulæ and anterior and posterior circumflex at the lower border of the subscapularis muscle.

Besides the large nervous trunks, branches derived
from them are met with passing to the muscles covering
the several walls of the space. Thus, the anterior wall
is supplied by the loop formed round the first part of
the artery by the external and internal anterior thoracic
nerves, the inner wall by the posterior thoracic,.the posterior wall by the subscapular, the outer by the circumflex and musculo-cutaneous.

The *lymphatic ganglia* are very numerous, and lie,
some along the course of the vessels, and some along the
lower border of the pectoralis major. These ganglia receive the lymphatics of the upper extremity, back, and
posterior part of neck, the lateral lymphatics of the trunk,
those of the epigastrium and anterior part of thorax, and
mammary region.

In order to gain a topographical idea of the position
of this vessel, in the *first* part of its course, *i. e.*, between
the clavicle and the upper border of the pectoralis minor
muscle, it is best exposed from the front, by detaching
the clavicular attachment of the pectoralis major and
turning it down, when it will be found lying in a triangle (the *subclavicular*), which is bounded above by the
subclavius muscle, below by the pectoralis minor, and
internally by the thorax. Immediately beneath the detached portion of the pectoralis major is a dense fascia
(the *costo-coracoid membrane*), a prolongation of the deep
cervical, passing beneath the clavicle, and attached to
the coracoid process and upper ribs, enveloping the pectoralis minor and binding down the axillary vessels in
a sort of sheath. From the lower border of the muscle
this aponeurosis descends, to be attached to that covering

in the pectoralis major and that forming the floor of the
axillary space.   It assists in forming the *fold* of the
axilla, and has been termed the "suspensory ligament."
Perforating this membrane are seen the cephalic vein,
passing into the subclavian, the acromio-thoracic vessels,
and the external anterior thoracic nerve.   In *front* of
the vessel in this part of its course lie the clavicular
portion of pectoralis major, the subclavius muscle, costo-
coracoid membrane, and cephalic vein, and the loop from
the outer and inner cords, giving off the external and
internal anterior thoracic nerve; *externally*, the cords of
the brachial plexus; *internally*, the axillary vein; *posteriorly*, the first intercostal muscle, second serration of
serratus magnus, and the nerve of Bell.

The *second* part of the course of this vessel lies beneath
the crossing of the pectoralis minor, and by dividing
the remaining portion of the pectoralis major and the
pectoralis minor, the whole extent of the vessel will be
exposed.

In *front* of the second portion of the vessel lie the
pectorals and inner head of median nerve; *externally*,
the external cord of the brachial plexus; *internally*, the
axillary vein and inner cord of the plexus; *posteriorly*,
the posterior cord of the plexus and the subscapularis
muscle, separated from it by a cellular interval.

The *third* portion of the vessel is beyond the pectoralis minor, and between it and the lower border of the
pectoralis major by which it is covered.   In *front* of
the third portion of its course lies the pectoralis major;
externally, the coraco-brachialis, the median, and musculo-cutaneous nerves; *internally*, the ulnar and internal cutaneous nerves and axillary vein; *posteriorly*, the
musculo-spiral and circumflex nerves, and the tendons

of the latissimus dorsi, teres major, and subscapularis muscles.

Ligature of the Axillary Artery.—The axillary artery may be tied in the first part of its course or in the third.

In the first part it is very deep and difficult of access, but it may be reached, either by separating the fibres of the pectoralis major and deltoid, or by means of a semi-lunar incision through the integument, extending from a little external to the sterno-clavicular joint towards the coracoid process, taking care to avoid the cephalic vein; next the clavicular fibres of the pectoralis major must be divided, the arm adducted, and the pectoralis minor drawn down. The costo-coracoid membrane, which is next seen on the stretch, must be opened, the cords of the plexus drawn outwards, and the axillary vein inwards, when the ligature can be passed from *within outwards*.

This operation is very difficult and dangerous, from the close relation of the axillary and cephalic veins, and the acromio-thoracic vessels.

In the third part the vessel is easily reached.

The arm is to be extended and supinated, in order to throw out the fold of the coraco-brachialis muscle, the inner border of which is the guide to the artery, and an incision of about three inches in length is to be made through the integument, rather nearer the anterior than the posterior fold of the axilla, the deep fascia being scratched through and the basilic or axillary vein avoided; the artery is seen lying crossed by the median nerve, and having the axillary vein to its inner side, and sometimes on it; these structures must be carefully isolated, and the needle passed from *within outwards*. Occasionally the vessel divides high up into the brachial and radial, thus complicating the operation of ligature. Again, after division, these trunks may reunite by cross branches, and the circulation continue as freely as before, unless a ligature be applied to each.

FIG. 21.

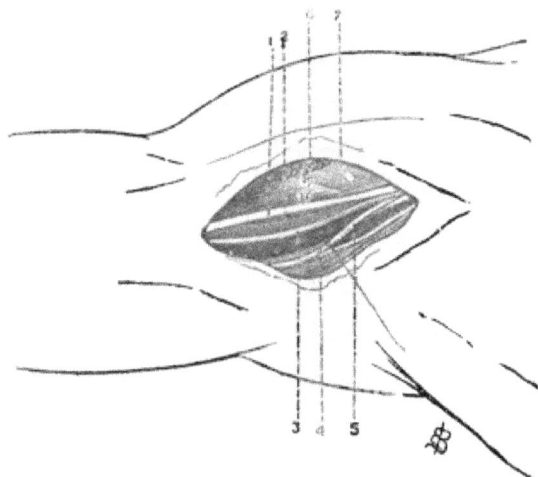

1. Median nerve. 2. Axillary artery. 3. Internal cutaneous nerve. 4. Axillary vein. 5. Ulnar nerve. 6. Coraco-brachialis muscle. 7. Deltoid muscle.

Collateral Circulation after Ligature of the Axillary.— Ligature of the axillary artery in the upper part of the first portion of its course, above the giving off of the acromial thoracic, may be regarded as equivalent to ligature of the subclavian in the *third part* of its course (*vide* Ligature of Subclavian). If the vessel be tied below this point, the subscapular inosculating with the suprascapular and posterior scapular, and the long thoracic with the internal mammary and intercostals, are called upon to restore the circulation. If the ligature be applied below the giving off of the subscapular, the posterior circumflex, anastomosing with the supra-

scapular and acromio-thoracic, and the subscapular with
the superior profunda, form the chief collateral channels.

SURGICAL ANATOMY OF THE BRACHIAL REGION

Surface Markings.—This region may be considered
as lying between a line drawn round the lower border
of the axilla, and another round the arm just above the
condyles of the humerus. Its general form is that of a
cylinder, flattened internally and externally, convex in
front, owing to the swell of the biceps muscle. Along
the inner aspect of the biceps is a well-marked groove,
extending from the axilla to the bend of the elbow;
whilst a shallow groove exists also on its outer surface,
becoming lost at the point of insertion of the deltoid;
beneath the skin is seen in the internal groove the
basilic vein, and external to the biceps, the cephalic.
The posterior surface of the brachial region is rounded
at about its middle, where the greater mass of the triceps muscle lies, and below this point its flattened
sharp-edged tendon forms towards its insertion into the
olecranon. In the inner groove lies the brachial artery,
where it can be felt or seen pulsating, overlapped by the
inner border of the biceps at about its middle, the muscle forming the guide to it throughout its course, either
for its deligation or compression. The surgical considerations affecting this region refer chiefly to amputation, ligature of its vessel, fractures, and the removal of
tumors.

On removing the skin, the subcutaneous cellular
tissue, divided into two laminæ by a layer of fat, is
first met with, and it is in this structure that the superficial nerves, veins, and lymphatics lie. The brachial

aponeurosis completely envelops the arm, and is thickest
at the back and sides; it is continuous above with that
covering the deltoid and subclavicular region, and below
with that covering the forearm, and sending processes
between the muscles forms septa, which are attached to
the humerus. The attachment of this aponeurosis is
very evident laterally, where it is inserted into the
condyles and condyloid ridges of the humerus, dividing
the region into two distinct compartments, an anterior
and a posterior. The *anterior* contains, successively,
the biceps, immediately beneath which, in the lower
half of the arm, is the brachialis anticus, inclosing at
the upper part of its origin, the insertion of the deltoid,
and passing downwards to its own insertion, covers
in the humerus completely between the lateral margin
of the anterior and posterior compartments. Behind,
and internal to the biceps above, is the coraco-brachialis; below and externally are, the musculo-spiral nerve,
the origins of the supinator longus and extensor carpi
radialis longior.

Lying along the inner border of the biceps is the
vasculo-nervous cord, formed by the brachial artery, its
veins, and the median, ulnar, external, and internal
cutaneous nerves; the median accompanies the artery
throughout, lying first outside it, then upon it, and lastly
internal. In the upper third, this vasculo-nervous cord
lies along the inner border of the coraco-brachialis, having
the long head of the triceps behind, and just on the
humerus, against which the vessel is easily compressed;
in its lower two-thirds it lies on the brachialis anticus.

The *posterior* aponeurotic compartment contains the
triceps, which covers in the entire posterior surface of
the humerus and the musculo-spiral nerve before it passes

anteriorly, as it lies in the musculo-spiral groove with
the superior profunda vessels, before it perforates the
septum to pass into the anterior compartment.   It also

Fig. 22.

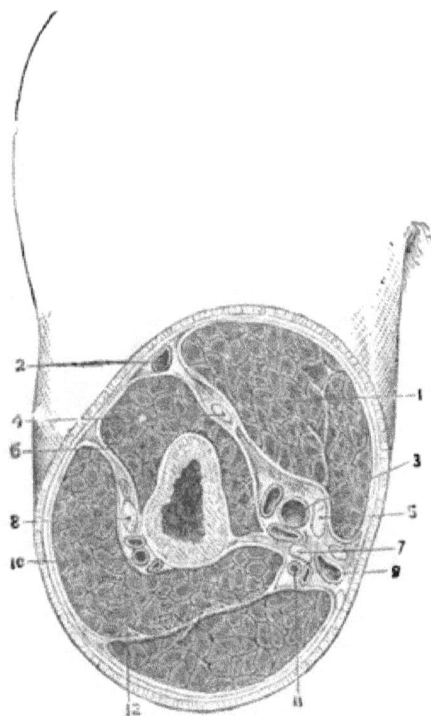

A section through the middle of the right upper arm. 1. Biceps.   2. Cephalic
vein. 3. Brachial vessels. 4. Musculo-cutaneous nerve. 5. Median nerve. 6.
Brachialis anticus. 7. Ulnar nerve. 8. Musculo-spiral nerve. 9. Basilic vein,
with internal cutaneous nerves. 10. Superior profunda vessels. 11. Inferior
profunda vessels. 12. Triceps, with fibrous intersection. (HEATH.)

contains the ulnar nerve, which at first lies in the anterior compartment, in contact with the brachial artery
above.

A correct knowledge of the relations of the layers of
these muscles, their aponeuroses, and the course taken

by the main vessel and its branches, determines the contents of the flaps in amputation of the upper arm.

Relations of Brachial Artery (above bend of elbow).— Its course is indicated by a line drawn from the junction of the anterior with the posterior two-thirds of the axilla, to the centre of the bend of the elbow. It is accompanied by venæ comites, which frequently interlace and conceal the vessel when cut down upon. In *front* of the vessel is the integument and fasciæ and the median nerve; *externally*, median nerve (in upper half), coraco-brachialis, and biceps; *internally*, internal cutaneous and ulnar nerves, and in lower half the median nerve ; *posteriorly*, lie the brachialis anticus, coraco-brachialis, musculo-spiral nerve, and superior profunda vessels, separating the middle and long heads of the triceps.

Ligature in the upper third is performed by making an incision about two inches in length along the inner border of the coraco-brachialis muscle; the subcutaneous tissue and aponeurosis are to be divided, taking care to avoid the basilic vein ; then the internal cutaneous and ulnar nerves will be found on the inner side of the artery, the median externally, and a number of venæ comites superficial to, and on each side of it. The needle should be applied from *within outwards*. Occasionally two vessels are found lying parallel to each other, or placed one over the other, the posterior lying very deep, the result of a high division ; under such circumstances it is obvious that it must be determined by pressure whether one or both communicate with the aneurism or wound.

In the middle third the vessel is not so easy to tie as might be imagined from its superficial position ; it is beneath a very dense part of the fascia, often overlapped by the biceps, and very movable beneath the integument.

It is reached by an incision along the inner border of this muscle, the forearm being slightly flexed so as to relax the biceps. The median nerve is usually first met with, crossing the vessel; this should be drawn inwards and the biceps outwards, and the venæ comites carefully separated from the trunk before a ligature is applied. In this instance, as in several others, it is advisable to make the incision slightly *oblique* to the real course of the vessel, as room is gained, and also a better view of any deviation from its natural course. The inferior profunda, if large, or the ulnar nerve, may sometimes be mistaken for this vessel.

Branches of the Brachial Artery.—The *superior profunda* is given off generally opposite the lower border of the teres major, and accompanies the musculo-spiral nerve; the *nutritious* artery enters the nutrient canal near the insertion of the coraco-brachialis, and passes downwards towards the elbow-joint; the *inferior profunda* accompanies the ulnar nerve; the *anastomotica magna* arises on the inner side, about two inches above the elbow-joint, and lies between the brachialis anticus and median nerve, crossing the latter, and finally divides into two large branches. In applying a ligature to the lower portion of the main trunk, the position of this vessel is of importance, as it is the chief means of carrying on the collateral circulation.

Collateral Circulation.—After ligature *above* the origin of the profunda superior, the posterior circumflex and subscapular, anastomosing with the ascending branches of the profunda, carry on the circulation; if *below,* the superior and inferior profunda, inosculating with the recurrent branches of the radial, ulnar, and interosseous, would maintain it.

The *action of the muscles* attached to the humerus upon the fragments in *fractures* of this bone, are generally as follows : If below the insertion of the muscles attached to the bicipital groove, but above the insertion of the deltoid, this muscle drags the lower fragment upwards and outwards, whilst the former set draw it towards the trunk. If below the insertion of the deltoid, the action of the muscles depends upon the direction of the fracture; that is to say, whether it be oblique or transverse.

Bearing in mind that the period of union of the shaft and its epiphysis is about the thirteenth year, we have an important diagnostic point, in cases of difficulty, between fractures immediately above the condyles, separation of the epiphysis, and dislocation of both radius and ulna backwards. If the fracture exists, the crepitation and ready reduction of the bones to their normal position would distinguish it from dislocation. Again, if instead of being transverse, the fracture of the shaft be oblique, in a direction from above downwards and forwards, the triceps will draw the lower fragment upwards and backwards; if oblique in a contrary direction, the lower fragment would be drawn upwards and forwards by the brachialis anticus and biceps.

Owing to the attachment of the flexors and extensors of the wrist and fingers to the condyles, unless the forearm be also put up in an angled splint, there is great liability to ununited fracture, as any movement of the wrist would tend to displace or rotate the lower fragment.

SURGICAL ANATOMY OF THE REGION OF THE ELBOW.

Surface Markings.—This region includes the articulation of the elbow-joint, and its immediate relations present for superficial examination four surfaces. Supposing the forearm fixed at a right angle with the humerus, a hollow is formed in *front* by the muscles attached to the condyles, within which lies the tendon of the biceps, its margin sharp externally, and flattened internally where its fascia is felt passing to the muscular mass attached to the inner condyle, and below which can be seen or felt the pulsation of the brachial artery; beneath the skin are visible, particularly in thin persons, the superficial veins of the bend of the elbow. On the *outer* side can be felt the external condyle, and the head of the radius; *internally,* the inner condyle; whilst *posteriorly,* nearer the inner than the outer condyle, is the olecranon, with the insertion of the triceps tendon.

The mutual relations of these structures to each other should be carefully examined in every position of the normal arm, in extreme and intermediate flexion and extension, and in pronation and supination of the forearm, and in cases of injury compared with the opposite side. In complete extension, the olecranon is above the level of the condyles; in semiflexion, on the same level; and in flexion, at a right angle below the level of the condyles.

Superficial Dissection of Bend of Elbow: Anterior Aspect.—On reflecting the skin, which is thin, lax, and thrown into folds, beneath the subcutaneous cellular tissue lie the cutaneous nerves, and an important plexus of veins and lymphatics. The nerves are derived from the internal cutaneous, lesser internal cutaneous, median and musculo-cutaneous. The superficial veins are derived principally from the radial, ulnar, and median veins, and at the outer border of the biceps tendon the median joins the basilic, by means of a large oblique branch, the median basilic; and also the cephalic, by a similar branch; the median cephalic, directed along the lower border of the biceps. At the point of bifurcation, the aponeurosis of the forearm is perforated by a communication between these veins and those accompanying the brachial artery, also between the superficial and deep lymphatics. The aponeurosis of the bend of the elbow is a continuation of the brachial, and is strengthened internally by an offset from the tendon of the biceps (the bicipital fascia). It is this process which separates the median basilic vein from the brachial artery. It receives, moreover, posteriorly, a slip from the tendon of the triceps, is attached to both condyles, and has a further slip from the brachialis anticus tendon.

On reflecting the aponeurosis, the mass of muscles which form the boundaries of the space are brought into view. *Externally*, from above downwards, in order, are the supinator longus, extensor carpi radialis longior and brevior, covering the lower and external part of the humerus, and the external lateral ligament. *Internally*, the common tendon of origin of the flexors and pronators of the wrist and forearm above; and passing between these muscular masses are, externally, the tendon of the biceps; internally, the tendon of the brachialis anticus, covering in the anterior part of the articulation, part of the head of the radius, and the coronoid process. These muscles form two wedge-like troughs, in the outer of which lie the musculo-spiral nerve, superior profunda and radial **recurrent vessels**; in the inner, the brachial artery, **its venæ** comites, and the median nerve.

The brachial artery at the bend of the elbow lies in the middle of this space, at first superficial, and afterwards, before it divides, deep, and opposite the coronoid process forms the radial and ulnar, and has the following relations: In *front*, integument and superficial fascia, median basilic vein, bicipital fascia; *externally*, tendon of biceps; *internally*, median nerve; *posteriorly*, the brachialis anticus.

The *lymphatics* are both superficial and deep, the former lying upon the aponeurosis, and communicating with one or more large ganglia immediately over the anterior aspect of the inner condyle; these ganglia become enlarged in poisoned wounds of the finger, and indurated in constitutional syphilis. The deep ones follow the course of the arteries. This aspect of the region is of the highest surgical importance, as in the operation of bleeding from the median basilic vein, there is a chance of wounding the brachial artery; in the event of such an accident occurring, either an aneurismal varix, or varicose aneurism, or a diffuse or circumscribed traumatic aneurism, might be the result. Arterial hemorrhage from wounds in this region, can be generally controlled by extreme flexion of the arm, aided by a pad of lint or small roller. Owing to the great number of inosculating vessels lying round the joint, there is considerable chance of secondary hemorrhage.

The *posterior* aspect of the region of the elbow presents for examination the olecranon process of the ulna, between which and the integument is a bursa; external to the olecranon can be felt the articulation of the head of the radius with the capitellum; and internally a deep depression, between it and the projecting inner condyle, in which lies the ulnar nerve and posterior ulnar recurrent artery, passing between the two heads of the flexor carpi ulnaris. The position of this nerve must be borne in mind in resection of the elbow-joint, and it must be

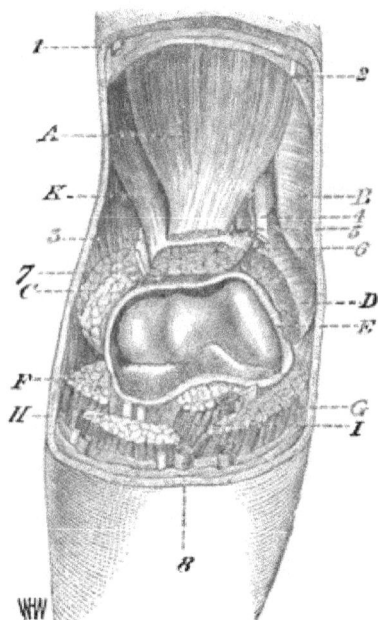

Fig. 23.

Structures in relation with the anterior aspect of the elbow-joint. 1. Cephalic vein. 2. Basilic vein and internal cutaneous nerve. 3. Musculo-spiral nerve. 4. Median nerve. 5. Brachial artery and venæ comites. 6. Anastomotica magna. 7. Radial recurrent. 8. Median vein. A. Biceps. B. Triceps. C. Supinator longus and extensor carpi radialis longior (the division between them is not evident enough). D. Origins of flexors and pronators. E. Capsule of joint. F. Extensor carpi radialis longior. G. Pronator teres. H. Supinator longus. I. Tendon of biceps (beneath which and the capsule is the brachialis anticus).

carefully isolated, and hooked out of the way in the subsequent steps of the operation, to avoid injury. The artery is only protected in this operation by the brachialis anticus, so that after the bones are divided, the knife must be used cautiously. The great vascularity of the region of the elbow-joint is, no doubt, one cause of the success attending its excision, the inosculating branches being the superior and inferior profundæ, anastomotica magna, anterior and posterior ulnar recurrent, interosseous recurrent, and radial recurrent. It is worthy of notice that the nutrient vessels of the humerus, radius, and ulna run *towards* the elbow-joint.

On making a vertical section through the elbow-joint, the following structures would be divided from above downwards: Skin and integument, containing the superficial veins and nerves already described, aponeurosis of arm, supinator longus, pronator teres, brachial artery and veins, and median nerve and musculo-spiral nerve, extensor carpi radialis longior, tendon of biceps, brachialis anticus, supinator brevis, flexor carpi radialis, extensor carpi radialis brevior, external and internal lateral ligaments, flexor sublimis, ulnar nerve, extensor carpi ulnaris, flexor carpi ulnaris, anconeus, olecranon, and olecranon bursa, and then the integument. The operation of amputation through the articulation is not one frequently performed, but the flaps would contain these structures.

The *articulation of the elbow-joint* admits of flexion and extension, and the direction of the articular surfaces not being parallel to a line drawn through the condyle, it follows that the axis of the forearm is not continuous with that of the arm, the trochlear surface being much lower down than the capitellum. The *ligaments* connecting the bones of the joint are very strong—namely, an anterior, a posterior, and two lateral, whilst there is a very large synovial membrane, the arrangement of which, in diseases of the articulation, causes the swelling to **take** place posteriorly **on** either side of the olecranon, and anteriorly at **the bottom** of the fold **at the** bend of the elbow.

Fractures of the humerus, in the **region of the** joint, frequently implicate it, **and are to be** distinguished from dislocation **by there not being the change in** the **normal** relation of *three* tuberosities—viz., the olecranon, epicondyle, **and epitrochlea.** Almost any form of dislocation may exist, but **the most common is that of** both **bones of the forearm backwards,** the strength **of the** lateral ligaments being **a great obstacle to** lateral displacement.

The *dislocations* to which the articulations of the elbow-joint are liable are, in the first place, of both **radius** and ulna, backwards, **outwards, inwards, forwards;** of the radius **only,** forwards and backwards; **and of the** ulna only, **backwards.** The difficulties **in the** diagnosis of these several conditions are frequently enhanced **from** the **complication with fracture or** separation **of the** epiphyses, the union of which is as follows: **the outer** condyle and both portions of the articulating portions **of** the humerus at the *sixteenth or seventeenth year,* the **inner** condyle at the *eighteenth* year, whilst the superior articular extremities of the **radius and ulna unite with their** shafts at *puberty.*

SURGICAL ANATOMY OF THE FOREARM.

Surface Markings.—The region **of** the **forearm, or** antebrachial, **may be** described as lying between the **lower margin of that** assigned to the elbow and the first fold seen before the wrist-joint. In shape it is that of a flattened **cone, which** varies in **form according as the**

limb is pronated or supinated, the roundness of its lateral
boundaries being due to the flexors and pronator teres on
the inner, and the supinators and extensors on the outer.
The bones, the radius and ulna, are capable of being
felt almost entirely throughout the region, particularly
the ulna, the posterior border of which is subcutaneous
from the olecranon to the styloid process; the radius,
however, has its shaft thickly covered with muscles, and
is felt with greater difficulty, and the several tendons of
the muscles clothing them are readily seen on putting
them in action. The radial and ulnar pulses are seen
in the lower part, while the radial can be felt in the
upper along the inner border of the mass of muscles at
the radial side.

Dissection.—On removing the skin, which is thin and
smooth, and provided with hairs, the subcutaneous fascia
is met with, containing a great deal of fat, in which lie
the superficial veins and cutaneous nerves and lymphatics. Beneath this layer is the antebrachial aponeurosis, continuous with that already described at the
elbow. It is dense, tough, and by its prolongation forms
fibrous sheaths for the muscles and other structures of
the region. This general antebrachial aponeurosis may
be considered as forming into two chief compartments,
attached laterally to the radius and ulna; thus the anterior is bounded by the anterior lamina, the anterior
surfaces of the radius, ulna, and interosseous membrane,
and the posterior by the posterior surface of these bones
and the interosseous membrane, and by the aponeurosis.

The anterior aspect of the forearm, according to this
arrangement, supposing the bones to be midway between
pronation and supination, contains from the surface to
the interosseous membrane, the integument of the forearm,

aponeurosis, the radial vessels and nerves, lying in a
cellular interspace between the radial and ulnar mass of
muscles, pronator teres as far as the middle third, palmaris longus, flexor carpi radialis and ulnaris, on the

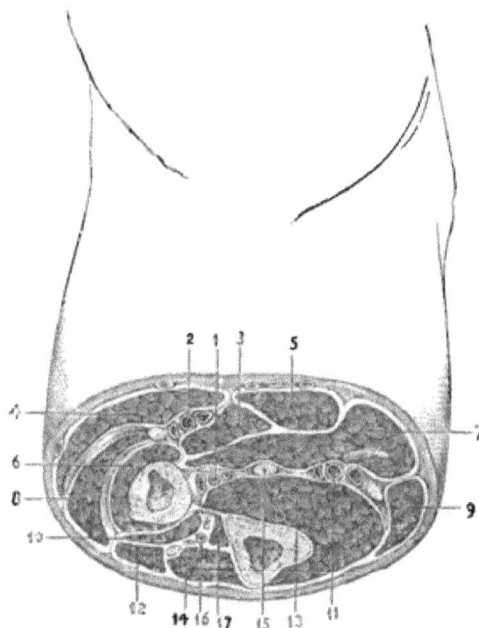

FIG. 24.

A section through the middle of the right forearm. 1. Anterior interosseous
vessels and nerve. 2. Radial vessels and nerve. 3. Pronator teres. 4. Supinator
longus. 5. Flexor carpi radialis. 6. Supinator brevis. 7. Flexor sublimis digitorum. 8. Extensores carpi radialis longior and brevior. 9. Flexor carpi ulnaris.
10. Extensor ossis metacarpi pollicis. 11. Ulnar vessels and nerve. 12. Extensor communis digitorum. 13. Flexor profundus digitorum. 14. Extensor carpi
ulnaris. 15. Median nerve. 16. Posterior interosseous vessels and nerve. 17.
Extensor secundi internodii pollicis. (HEATH.)

inner aspect, and the supinator longus, extensor carpi
radialis, longior and brevior, beneath which muscles is the
supinator brevis, the flexor sublimis digitorum, median
nerve, and ulnar vessels and nerve, flexor profundus, and
flexor longus pollicis, lying on the bones and inter-

osseous membrane. In the lower third, the tendons of the muscles passing to the wrist lie over the pronator quadratus, and upon the membrane itself the anterior interosseous vessels and nerves.

The posterior compartment contains from the surface to the bones, the extensor communis digitorum, slightly covered in above by the extensor carpi radialis longior, and lower down lying along the extensor carpi radialis brevior, extensor minimi digiti, extensor carpi ulnaris, and anconeus. The second layer comprises the extensor ossis metacarpi, primi and secundi internodii pollicis, the extensor indicis, and the posterior interosseous vessels and nerves. The extensors of the thumb will be seen to form a sort of spiral round the lower part of the radius; they are inclosed in a sheath which is very liable to be the seat of severe teno-synovitis, especially met with in reapers and workmen.

Relations of the Radial Artery in the Forearm.—Supposing the vessel to be normal, a line drawn from the middle of the bend of the elbow to the inner side of the styloid process of the radius represents its *course* in the upper third; it is concealed by the pronator teres and by the edge of the supinator longus, and lies on the tendon of the biceps, and in a bed of fat and cellular tissue, in relation with some muscular branches of the musculo-spiral nerve, and on the supinator brevis. In the middle third it lies between the tendons of the flexor carpi radialis and supinator longus, having the radial nerve to its outer side, and on the pronator teres and flat head of the flexor sublimis. In the inferior third it has the same tendons on either side, whilst it rests upon the flexor longus pollicis, pronator quadratus, and radius. It is accompanied throughout by venæ comites.

To tie the Radial Artery.—The vessel is readily secured throughout its course, and any incisions made through the integument along the margin of the supinator or its tendon would reach it, but its deligation on the upper third is only required in cases of wound or operation, which, in the former instance, would be enlarged, and the vessel tied above and below the point of injury. The vessel is liable to several irregularities, the most common of which is a high origin from the brachial when it lies often merely subcutaneous and can be seen pulsating throughout its course.

Relations of the Ulnar Artery in the Forearm.—Arising at the bifurcation of the brachial, it runs obliquely downwards, on the ulnar side of the arm towards the radial side of the pisiform bone, and in its upper third it is covered by the pronator teres, flexor carpi radialis, palmaris longus, and flexor sublimis, and having on its inner side the flexor carpi ulnaris, on its outer the flexor sublimis, and resting on the brachialis anticus. In the lower two-thirds, it is covered in by the fasciæ; on its inner side are the flexor carpi ulnaris and ulnar nerve, on its outer the flexor sublimis, and it rests on the flexor profundus; venæ comites accompany it throughout. This vessel occasionally arises high up, when it also lies immediately beneath the integument and fasciæ, uncovered by muscles. The same remarks, with regard to tying it in the upper third of the arm, may be made as have been in the case of the radial. It is very difficult to secure in its upper third, owing to being so covered in by muscles.

The *nerves* of the forearm are cutaneous and deep, the former being branches of the external and internal cutaneous as far as the wrist, of the lesser internal cutaneous and musculo-spiral, and in the lower third of the ulnar and radial.

The musculo-spiral enters the forearm between the brachialis anticus and supinator longus, and after breaking up in the supinator brevis is distributed to the extensors and supinators; the median enters the forearm between the two heads of the pronator teres and is accompanied by a vessel, the comes brevis mediani, which is occasionally of considerable size, and may form one of the chief supplies to the palm; the ulnar enters the forearm between the two heads of the flexor carpi ulnaris, and accompanies the artery of the same name.

In amputation through the upper third of the forearm the flaps would contain the structures as follows: In the *anterior,* integument and superficial veins and nerves, the flexor carpi radialis, supinator longus, palmaris longus, flexor carpi ulnaris, extensor carpi radialis longior, pronator teres, flexor sublimis and profundus, and the radial, ulnar, and anterior interosseous vessels and nerves. The *posterior,* the extensor carpi radialis brevior, supinator brevis, and the ulnar and common extensor of the fingers, with the posterior interosseous vessels and nerves, integuments, and superficial veins and nerves. The vessels requiring ligature are the radial and ulnar, which will be found just beneath the integument, the anterior and posterior interosseous, which retract along the interosseous membrane, and perhaps the comes nervi mediani.

In performing the flap amputation the bones should be placed midway between pronation and supination; and in cases of fracture of one or both bones, the same position must be maintained, as they are then most nearly parallel and furthest separated from each other, and there is less chance of any union between them. Moreover, owing to the tendency to motion that exists between these bones, a false joint may be the result unless accurate adaptation and perfect rest be maintained.

The fractures which occur in the forearm are those affecting either one or both bones and their processes. If the coronoid process be fractured, a rare accident, dependent on muscular action or dislocation, that amount of flexion performed by the brachialis anticus is necessarily lost; symptoms of fracture of the olecranon are obvious, and the powerful action of the triceps in separating the fragments accounts for the rarity of bony union; moreover, the articulation is generally opened. If the neck of the radius be fractured, a result of direct violence, the diagnosis is obscure, and must be deduced from the want of power of *voluntary* pronation and supination.

The *action of the muscles* of the forearm upon fragments is well marked in such a case as fracture of the shaft of the radius alone. Thus, supposing it broken about its centre, the upper fragment is drawn forwards by the biceps, inwards by the pronator teres; the lower fragment is pronated and drawn downwards and inwards by the pronator quadratus, and its styloid process tilted upwards by the supinator longus. In fractures of both bones the action of the muscles often causes great deformity.

SURGICAL ANATOMY OF THE REGION OF THE WRIST AND BACK OF HAND.

Surface Markings.—Beneath the integuments are seen the cutaneous veins and the tendons of the muscles acting on the wrist and fingers, the *anterior* ones being most evident in flexion of the hand or clenching the fist,

14

especially the palmaris longus and flexor carpi radialis;
the *posterior*, in extension of the wrist or fingers, and in
extension of the thumb.   The position of the styloid processes of the radius and ulna can be felt, the former
lower down than the latter; the relation of these processes, however, is altered during pronation and supination.   In front of the styloid process of the radius is the
root of the thumb and prominence of the scaphoid, and
on the inner side is the pisiform bone, with the flexor
carpi ulnaris attached to it.   The pulsations of the
radial and ulnar arteries can be easily felt and generally seen; the former on the radial side of the flexor
carpi radialis; the latter on the radial side of the flexor
carpi ulnaris.

Dissection.—On reflecting the skin, the subcutaneous
cellular tissue is seen, free from fat, and lying in it are
the cutaneous vessels, nerves, and lymphatics.   The
aponeurosis is a continuation of that forming the sheath
of the muscles of the forearm, which is remarkable at
the wrist as being very strong, and affording special
channels for the passage of the tendons of the muscles
of the forearm.   It consists of two portions—an anterior, very thick and strong (anterior annular ligament),
continuous with the deep fascia of the forearm, which is
attached to the pisiform, unciform, radius, scaphoid and
trapezium bones, receiving an expansion of the tendon of
the palmaris longus, and forming an arch between the
thenar and hypothenar regions; and a posterior, dense,
and formed of circular and longitudinal bundles of tissue
attached to the ulna, cuneiform, pisiform bones, to the
radius, and to the several eminences on its dorsal aspect,
which separate the extensor tendons, and is thus divided

into six compartments, each lined with a separate synovial membrane.

Anterior Region of Wrist.—The first layer of structures beneath the integument and aponeurosis of the
wrist, and described from the radial towards the ulnar
side, consists of—the supinator longus, the radial vessels, the flexor carpi radialis and palmaris longus, the
former of which perforates the denser portion of the annular ligament, the ulnar vessels and nerve, and the
flexor carpi ulnaris. All these tendons have separate
sheaths, derived from the annular ligament and synovial
membranes. The second layer consists of the flexor
sublimis, enveloped in synovial membrane, and the
median nerve, with its accompanying artery. The third
consists of the flexor longus pollicis and the flexor profundus digitorum, having their synovial sheaths in
common with the superficial flexor, while close on the
bone are the carpal branches of the radial and ulnar.

The Posterior Region.—Between the skin of the back
of the hand and the subcutaneous tissue are a number of
superficial veins and cutaneous nerves, derived from the
radial and dorsal branch of the ulnar. Beneath them is
a strong fibrous membrane, apparently continuous with
the dorsal portion of the annular ligament. There is a
deeper layer, covering the bones and interossei, which is
blended with the palmar aponeurosis laterally, and the
dorsal tendons pass between these layers. Beneath the
annular ligament are the six compartments for the following tendons, beginning on the radial side: (1) extensores ossis metacarpi and primi internodii pollicis; (2)
extensores carpi radialis, longior, and brevior; (3) extensor secundi internodii pollicis, crossing the two former
very obliquely; (4) extensores digitorum, and indicis;

(5) extensor minimi digiti; (6) extensor carpi ulnaris.
The synovial membrane investing their anterior surfaces
being very thin and indistinct, is more frequently the
seat of *ganglion* than that of the flexor tendons.

The tendons of the extensor muscles, having arrived
at the metacarpo-phalangeal articulation, receive the
tendons of the lumbricales and interossei, whilst at the
first phalangeal joint they divide into three fasciculi, the
central one being inserted into the base of the second
phalanx, and the two lateral passing on and reuniting,
are inserted into the base of the ungual. They have no
distinct synovial sheaths.

The radial artery at the wrist can be felt or seen beat-
ing between the tendons of the flexor carpi radialis and
supinator longus, where it is quite superficial and easily
secured. Accompanied by venæ comites, it winds round
the outside of the wrist, to gain the first interosseous
space, when it enters the palm between the two heads of
the first dorsal interosseous, and is crossed by the extensors of the thumb. It is readily secured at the base of
the well-marked hollow formed by these muscles. In
disarticulation of the metacarpal bone of the thumb it
stands a chance of being divided, but if the knife be kept
close to the bone it can be avoided. The most important
branch of the radial is the superficial volar, which ordi-
narily lies subcutaneously, and completes the superficial
arch. The other branches given off at the wrist supply
the carpus and dorsal aspects of the thumb and first
finger.

The ulnar artery, at the wrist, lies with its venæ
comites on the radial side of the flexor carpi ulnaris, and
with its nerve to its inner side.

Articulations of the Wrist-joint.—These are the inferior

radio-ulnar, the radio-carpal, which exist between the lower end of the radius, the scaphoid, and semilunar bones. The synovial membrane of this articulation is also extended over the cuneiform bone and the inter-articular fibro-cartilage between the ulna and that bone; the intercarpal and carpo-metacarpal joints, which include the anterior articular surfaces of the cuneiform, semilunar, and scaphoid, the entire unciform, os magnum, and trapezoid, with the bases of the four inner metacarpals, have a common synovial membrane, whilst the pisiform and the trapezium have one each. Of these articulations the most important to the surgeon is the radio-carpal, as dislocation of the hand and carpus from the radius, either backwards or forwards, takes place here. Moreover, amputation is occasionally performed at this joint.

Fracture of the radius just above the articulation (Colles's fracture), is almost always transverse, and in young subjects is a separation of the lower epiphyses. The deformity is well marked; the result of the combined action of the supinator longus, extensors of thumb, and radial extensors of the wrist, causes the lower fragment to make a partial rotation on its transverse axis.

In examining the lower end of the forearm, in cases of injury, it must be remembered that the *head* of the ulna is prominent in *pronation*, and its *styloid* process in *supination*, owing to the rotation of the radius at its inferior radio-ulnar articulation.

Amputation at the wrist-joint may be performed either by a semilunar dorsal flap, and an anterior, formed from the palm, or by rectangular flaps. The knife may get notched against the pisiform bone, so that some little neatness is necessary in avoiding it. The styloid pro-

cesses of the radius and ulna require removal, and the
vessels which would be ligatured are the superficialis
volæ, some branches of the ulnar in the palm, and perhaps an abnormal median.

For the operation of *excision* of the joint, an accurate
knowledge of its component parts and relations is of the
utmost importance, as this excision, like that of the
ankle, reduces itself to an anatomical problem—viz., to
remove the disease, and at the same time to preserve to
the hand the tendons passing from the arm to it, and if
possible to retain their functions. It is difficult to lay
down any distinct rules, as obviously each case has its
own peculiarities; but by far the most scientific proceeding is that of Professor Lister, for an account of
which somewhat complicated but most successful operation, reference must be made to special works on operative surgery.

SURGICAL ANATOMY OF THE PALM.

Surface Markings.—This region extends from the wrist
to the web of the fingers. It is concave, and presents
two muscular eminences: one on the radial side, due to
the mass of muscles acting on the thumb, called the
thenar prominence, and the other due to muscles of the
little finger, the *hypothenar*. The intermediate space is
marked by several furrows, indicating the more marked
flexions of the hand, one of which, the oblique central
one, lies almost in the course of the superficial palmar
arch. The deep palmar arch may be referred to the
surface marking, by an imaginary line drawn between
the centres of those circles, which form the bases of the
thenar and hypothenar eminences. The bifurcation of

the digital vessels takes place a little nearer the palm
than the web of the fingers, and their course is subsequently along the under lateral aspects of the digits.
The skin is very sensitive, notwithstanding the horny
condition it acquires in those who have much manual
labor, and it is furrowed by ridges, in which lie the
orifices of the sweat-ducts. The great vascularity of the
skin of the palm predisposes this region to the occur-
rence of erectile tumors.

Dissection.—The subcutaneous tissue is full of lobu-
lated fat; beneath the fat and cellular tissue is the palmar
fascia, particularly strong in its central fasciculus, and
into the posterior portion of which is inserted the palmaris longus; opposite the heads of the metacarpal bones
it divides into four slips, which slips themselves will be
seen to divide into two processes, attached to the sides
of the first phalanx, giving passage to the flexor tendons, whilst the intermediate spaces transmit the digital
vessels and nerves.   Vertical septa pass down and divide
the central set of palmar muscles from the thenar and
hypothenar, the expansion of the palmar fascia covering
which is very thin.   This palmar aponeurosis is frequently the seat of contraction which implicates the
fingers.   The little palmaris brevis muscle is attached to
the skin and ulnar aspect of the central portion of the
aponeurosis.   On removing the palmar fascia the underlying structures are met with in the following order,
dissecting down to the metacarpus.

Commencing with the *thenar* eminence beneath the
fascial covering, lie the superficialis volæ artery, abductor pollicis, opponens pollicis, and radial head of short
flexor of thumb, tendon of flexor longus pollicis, ulnar
head of flexor brevis pollicis, princeps pollicis, and

radialis indicis arteries, metacarpal bone of thumb, trapezium, and tendon of flexor carpi radialis.

In the middle segment of the palm, beneath the central
fasciculus of the palmar fascia and the anterior annular
ligament, with which the fascia is continuous, lies the
superficial palmar arch and its digital branches, the ulnar and median nerves with their digital branches, the
tendons of the flexor sublimis and profundus digitorum,
with which latter are associated the lumbricales (these
muscles inclosed in their synovial sheaths), next a layer
of fibrous tissue separating them from the deep arch, the
deep branch of the ulnar artery, the adductor of thumb
and interossei, whose tendons with those of the lumbricales pass into the general dorsal aponeurosis, and lastly,
the metacarpus.

Beneath the palmar fascia of the *hypothenar* eminence
lie the palmaris brevis, some cutaneous vessels and nerves,
the abductor and flexor brevis, minimi digiti, commencement of superficial palmar arch, with its accompanying
ulnar nerve, opponens minimi digiti, deep branch of ulnar artery and nerve, and fifth metacarpal bone.

The metacarpal bone of the thumb articulates with
the trapezium by a saddle-shaped surface, and its shaft
is considerably curved anteriorly; and in amputation at
the metacarpo-trapezial joint, the secret of enucleating
the bone neatly consists in abducting it forcibly and dividing the lateral ligaments, of which it is better to cut
the inner one first. There are several methods of performing this operation, but that should always be chosen
which will leave the greatest amount of *opposing* tissue;
as the muscular pad, resulting from the flaps, even
though it lose its bony support, is of great importance
when the hypothenar mass of muscle is intact, as it will

in great measure retain the power of approximation.
Care must be taken to avoid wounding the trunk of the
radial artery as it passes between the two heads of the
first dorsal interosseous muscle, if possible. The metacarpal bones of the fingers, having a common synovial
membrane with their carpal bones, ought, if possible, to
be removed without disarticulation, owing to the liability of general suppuration. There is some little difficulty attending amputation of the fifth metacarpal bone,
owing to its double articulation with the os cuneiforme.

Abscess in the palm (*palmar abscess*), unless opened
early, is liable to spread up the arm, along the synovial
sheaths of the muscles, by passing beneath the annular
ligament, the excruciating pain attending it being due
to the tenseness of the palmar fascia. In opening collections of pus in the palm, the position of the palmar
arch must be recollected, and the knife should be entered
upon the head or neck of the metacarpal bone, and not
between the fingers, so that the bifurcation of the digital
artery may be avoided.

In wounds of the palmar arch, if ligature of the radial
and ulnar fail, the circulation is probably carried on by
an enlarged anterior interosseous or comes nervi mediani, and ligature of the brachial must be had recourse to.

Bursal tumors are commonly met with in association
with the synovial sheaths of the flexor tendons in the
palm, and generally communicate beneath the annular
ligament, with the continuation of these sheaths in the
forearm.

In the fingers the skin is very thick, particularly on
the palmar aspect, highly vascular, and freely supplied
with nerve-fibres. The subcutaneous tissue contains a
good deal of fat; beneath this tissue is the sheath or

theca of the flexor tendons, an osseo-fibrous canal, formed by the phalanges and a dense tube of fibrous tissue, disposed in circular and oblique bands, very thin immediately opposite the flexures, and perforated at the roots of the fingers for the passage of vessels and cellular tissue. The sheath is very thin on the palmar aspect of the ungual phalanx, and purulent infiltration into it is common at this point. The flexor tendons which lie in the sheath, are those of the flexor sublimis (*perforatus*) attached by two slips to the sides of the second phalanges, and those of the flexor profundus (*perforans*) which divides them, and is inserted into the base of the ungual. The canal is lined with a synovial membrane, which is reflected on to the tendons. Slender tendinous filaments, called *vincula*, connect these tendons to the walls of the canal. On the dorsum there is a strong aponeurosis formed by the extensor tendons, further strengthened by the expansion of the interossei and lumbricales. The common extensor passes on to the second phalangeal articulation, opposite which it divides into three fasciculi, the central one being inserted into the base of the second phalanx, whilst the two lateral slips reunite and pass on, to be inserted into the base of the ungual.

In *excision* of the phalangeal articulations it is necessary to retain as much as possible of this dorsal aponeurosis, in order that the power of extension may be afterwards kept. In *disarticulation* of a phalanx, it must be remembered that immediately the lateral ligament is divided the joint is opened, and that this lateral ligament does not coincide with the palmar fold of the digit, but is a little in front of it, so that the guide for entering the knife is the second fold in the skin seen on the dorsal aspect of the joint.

The *vessels* and *nerves* supplying the fingers, including the thumb, are derived from the digital branches of the radial and ulnar arteries, and ulnar and median nerves, the former supplying the thumb and radial side of the index, the ulnar side of the index, and remaining fingers; and the latter the little finger, half the ring finger, and the remaining fingers. The arteries anastomose very freely in the pulp and matrix of nail. The veins accompany the arteries and pass posteriorly, and form a considerable plexus, passing along the back of the fingers into the veins of the back of the hand.

Dislocation of the first phalanx of the thumb is difficult of reduction, and if reduced, of being maintained in position, owing to the great power exerted by the mass of muscles forming the ball of the thumb, and to the fact that when the phalanx lies on the dorsal aspect of the metacarpal bone, the narrow head of the latter becomes constricted by the two terminal attachments of the flexor brevis pollicis, the tendons of which are further strengthened by the conjoined insertion of the adductor and abductor.

CHAPTER V.

SURGICAL ANATOMY OF THE ABDOMEN.

BEFORE commencing the strictly surgical anatomy of the abdomen, it will be convenient to point out generally, the relations of the contents of this cavity to the surface of the body, considerations of importance to the surgeon as aids to diagnosis.

Relations of the Viscera to the Abdominal Parietes.— In order to facilitate the description of the abdomen, it can be mapped out by certain arbitrary lines into nine regions, to which the contained viscera can be referred —a method of reference of considerable use in a certain way, but of no great value as bearing upon the subject of Surgical Anatomy, properly so called. These lines are vertical and horizontal; the vertical, passing from the seventh costal cartilage to the middle of Poupart's ligament, on either side, and the horizontal through the level of the ninth costal cartilages, and crests of the ilia. The regions thus indicated are called the—

Right hypochondriac.	Epigastric.	Left hypochondriac.
Right lumbar.	Umbilical.	Left lumbar.
Right iliac.	Hypogastric.	Left iliac.

The *stomach* lies in the epigastric and left hypochondriac regions; the *transverse colon* crosses the umbilical; the *jejunum* and *ileum* occupy the umbilical and hypogastric; the *ascending* and *descending colon* the iliac and lumbar of either side; and the *kidneys* lie in the right and left lumbar, *the spleen*, the left hypochondriac. The student is frequently called upon to map out or percuss some larger viscus—such for example, as the *liver*—the position of which is indicated as follows: It fills the right hypochondriac region and concavity of the diaphragm, and is almost completely concealed by the overhanging of the ribs; part of the left lobe lies in the epigastric and left hypochondriac regions. It projects upwards into the thorax, and is separated from its wall by the thin lower margin of the right lung and the diaphragm. In this region its upper margin is about the fifth intercostal space; referred to a line drawn from the posterior boundary of the axilla vertically to the crest of the ilium, its margin would be about the seventh intercostal space. In the mesial line it is not easy to distinguish the upper boundary of the liver from the lower margin of the heart.

For *surgical* and *operative* reference, however, the regions of the abdomen will be considered under the heads of—(1) Abdominal parietes; (2) Such portions of the abdominal cavity as are of practical surgical importance; (3) Pelvis.

(1) *Abdominal Parietes.*

The region of the abdominal parietes may be divided as follows—(1) Antero-lateral; (2) Inguinal; (3) Crural; (4) Lumbar.

(1) *Antero-lateral Region of the Abdomen—Dissection.*— On removing the skin, which is very lax, excepting at the umbilical depression, is the subcutaneous cellular tissue and fascia, which latter may be divided into as

many lamellæ as the skill of the dissector permits of; two
of which, however, may be considered as ample for all
surgical purposes, the superficial and the deep. These
fasciæ are continuous, and interlace on either side of the
linea alba, which is seen as a well-marked depression,
and some of their fibres are attached to it, and in the
lower portion of the abdomen pass downwards to the
scrotum, strengthening the suspensory ligament of the
penis, and giving origin to the dartos (*vide* Inguinal
Region). In the fat existing between these laminæ lie
numerous cutaneous branches of the superficial epigastric and intercostal vessels and nerves. Beneath the
deeper layer is the external oblique muscle, the fibres of
origin of which interdigitate with those of the serratus
magnus and latissimus dorsi, and its aponeurosis passes
inwards, to be inserted into the iliac spine, the linea
alba, and iliac and pubic crests. The fibres of this
muscle are oblique from above downwards. Beneath it
lies the internal oblique, separated from it by a thin
cellular interval, derived originally from the lumbar
aponeurosis, and in which lie filaments of the lower intercostal and last dorsal nerves; its tendon is inserted
into the cartilages of the lower ribs, the linea alba, and
ilio-pectineal line, inclosing the rectus for its upper two-
thirds, and passing in front of it entirely for the remaining third. The fibres run in the contrary direction to
those of the former muscle. The transversalis is separated from it by a thin cellular interval, and having its
origin from the lower ribs and lumbar aponeurosis is inserted along the linea alba and by the conjoined tendon,
the upper three-fourths passing behind the rectus, and
blending with the tendon of the internal oblique, whilst
its lower fourth passes in front of it. It is separated

from the peritoneum by the transversalis fascia. The
rectus abdominis passes vertically down the abdominal
parietes, on either side of the linea alba, inclosed in a
partial sheath, and having three or four tendinous intersections, which are very readily seen beneath the skin
when the muscle is in action. Below it is a little muscle, the pyramidalis, attached below to the pubis and
inserted into the linea alba, at the junction of the middle
with the lower third of the rectus. The continuity of
the thin aponeurotic laminæ, which exist between the
abdominal muscles, with the lumbar aponeuroses, accounts for the occasional pointing of lumbar abscess in
the parietes above Poupart's ligament, or at the edge of
the rectus. The *deep epigastric* artery, with its veins,
after passing between the peritoneum and transversalis
fascia, gets behind it, and entering its fibres, freely
anastomoses with the superior epigastric from the internal mammary, which vessel itself enters the rectus below
the cartilages of the eighth or ninth ribs. Between the
transversalis fascia and the peritoneum is the subperitoneal cellular tissue. The peritoneum itself closely
lines this region of the abdominal wall, excepting for a
short distance above the pubes, where its attachment is
very lax—a circumstance taken advantage of in puncture of the bladder in this region.

Operations for the removal of abdominal tumors and
for paracentesis abdominis are performed in the mesial
line, as the simple structure of the linea alba, presenting no *layers*, does not admit of infiltration into muscular aponeuroses or sheaths. In the operation of paracentesis the parietal layer of the peritoneum only would
be involved; but in operations on an *ovarian cyst*, for
example, both parietal and visceral are divided. In

the case, however, of the incision through the parietes made for the application of ligatures to the iliac arteries, the abdominal muscles are divided, whilst the peritoneum is kept entire. The resilience of the walls of this region allows of the ready manipulation of its contents, with a view to diagnosis. Penetrating wounds are frequently followed by hernial protrusion, on account of the laxity of their cicatrices.

In the linea alba, between the two recti, is the *umbilicus*, the cicatrix of a fœtal structure, the umbilical cord, which having been cut or tied at birth, has shrunk up to its attachment at the abdominal parietes, and the closure of the opening is formed by tough fibrous tissue, closely adherent to the peritoneum and neighboring tissues.

Umbilical herniæ, if formed at an early period, possess a very thin covering of peritoneum, as that membrane is but recently formed at that spot, and in almost all cases this sac contains omentum, transverse colon, or small intestine.

Umbilical hernia is almost always a congenital affection; ventral hernia, though occurring at this spot, is really a protrusion through a preternatural opening in the fibrous coverings close to the umbilicus.

The regions of the abdominal parietes of greatest interest to the surgeon, and which demand a more accurate anatomical description, are those connected with inguinal and crural hernia; and though these protrusions are formed of structures contained in the abdominal cavity, they would be best considered in the relations in which they would present themselves—namely, from the surface.

SURGICAL ANATOMY OF THE INGUINAL REGION.

Surface Markings.—This region may be indicated by
a line drawn from the anterior superior spine of the
ilium to the mesial line of the body superiorly, the
mesial line itself, internally, and Poupart's ligament,
below. Between the bony prominences, the anterior
superior spine externally, and the crest of the pubes internally, is a curved furrow, with its convexity downwards, indicating the fold of the groin, and the position
of Poupart's ligament. Just above the pubes can be
felt the external abdominal ring, and the structures
forming the cord are readily recognizable.

Before commencing the dissection, the finger should
be passed up into the external ring without using any
force, by tucking up the scrotal tissue upwards along
the cord, in order to ascertain under what conditions of
the position of the leg the examination of the canal
could be most readily made, and the exact relations of
the structures composing it felt.

Dissection.—In order to arrive at a correct idea of
the parts concerned in inguinal hernia, it is advisable
to endeavor to obtain a simultaneous view of both surfaces of the abdominal parietes in this region, and for
this purpose a flap containing the entire thickness of
the walls should be made (including the umbilicus), so
that in prosecuting the dissection the actual relation
of the structures both at the commencement and termination of the course taken by a hernia may be examined
with facility.

On reflecting the skin, the superficial fascia is readily
divisible into two layers, between which lie the cutaneous vessels and lymphatics, the deep layer being closely

attached to Poupart's ligament and to the crest of the ilium. This fascia is freely movable over the subjacent

FIG. 25.

Superficial dissection of the inguinal and femoral regions. *a.* Superficial layer of fascia (reflected). *b.* Deeper layer of fascia (reflected), the superficial vessels being left attached to the external oblique. *c.* Inguinal lymphatic glands. *d.* Superficial circumflex iliac artery. *e.* Superficial epigastric artery. *f.* Superior external pudic artery. *g.* Poupart's ligament. *h.* Intercolumnar fascia. *i.* External abdominal ring. *k.* Arciform fibres of external oblique. *l.* Internal saphena vein. *m.* Femoral lymphatic glands. *n.* Ilio-inguinal nerve. *o.* Saphenous opening. (From Wood, on Rupture.)

aponeurosis, a circumstance which is taken advantage of by the surgeon in making his incisions down upon a strangulated hernia, when, owing to this laxity, by pinching up a fold of integument and transfixing it, he obtains a linear incision, which does not include either sac or gut. It is, moreover, of great value for the introduction of the needles in Wood's operation for the radical cure. Between these layers of fascia are the lymphatics, transmitting the ducts from the external genitals, the termination of ilio-hypogastric and ilio-inguinal nerves, and three superficial branches of the common femoral artery—viz., the superficial epigastric, superficial circumflex iliac, and superficial external pudic, with their accompanying veins. In plastic operations for the relief of extroversion of the bladder, the superficial epigastric should be carefully preserved to nourish the flaps. Beneath the deep layer of the superficial fascia is the aponeurosis of the external oblique; its lower portion, by its union with the fascia lata and deep fascia, forms the *crural arch*, which extends from the anterior superior spine of the ilium to the spine of the pubes, having also an attachment to the ilio-pectineal line (Gimbernat's ligament). The attachment to the ilio-pectineal line is strengthened by a triangular band of fibres passing upwards and inwards towards the linea alba, behind the inner pillar of the external ring. It will be noticed that extension and abduction of the leg renders the crural arch tense, so that in the examination of the parts or in the application of the *taxis*, the external abdominal ring must be relaxed by flexion and abduction. The two *pillars* of the external ring are bound together by a set of aponeurotic fibres, which interlace, more or less, over the whole of the inner part of this region, constituting the *intercolumnar bands*, from which a fascia is derived, forming a covering to the emerging cord—the *intercolumnar* or *external spermatic* fascia. It is the weakness or giving way of these bands which favors the hernial protrusion. The inner pillar of the ring is flat and riband-shaped, whilst the outer is sickle-shaped and thick, and upon it the cord or round ligament rests. On detaching a "dog's ear" of the aponeurosis of the external oblique along Poupart's ligament, immediately beneath it is a cellular interval separating it from the muscular fibres of the internal oblique, and the *conjoined tendon* of this muscle and the transversalis, this latter passing in front of the rectus and pyramidalis to the linea alba and pubes; blended with the lower fibres of the internal oblique and transversalis are the fibres of the cremaster muscle, on which the ilio-inguinal nerve lies. On carefully detaching the muscular fibres of the internal oblique from Poupart's ligament, and reflecting them outwards and inwards, the fibres of the transversalis are met with, forming an arch over the cord, and beneath this arch is the spout-like prolongation of the transversalis fascia (*the infundibuliform*) investing it. Behind the transversalis and the rectus, is the transversalis fascia, closely lining them, and here forming with the subperitoneal aponeurosis the posterior layer of the sheath of the latter muscle; in its lower fourth it is attached to the under surface of the crural arch, becoming continuous with the fascia iliaca. Beneath this fascia is the parietal layer of the peritoneum.

The position of the *deep epigastric vessels* can be easily seen, lying beneath the transversalis fascia and the peritoneum, and passing obliquely upwards and inwards, to gain the under surface of the rectus at about its lower third and internal to the cord.

It will be now found convenient to turn down the flap consisting of the entire thickness of the abdominal wall

as suggested, in order to obtain a view of these structures
from the peritoneal surface.

FIG. 26.

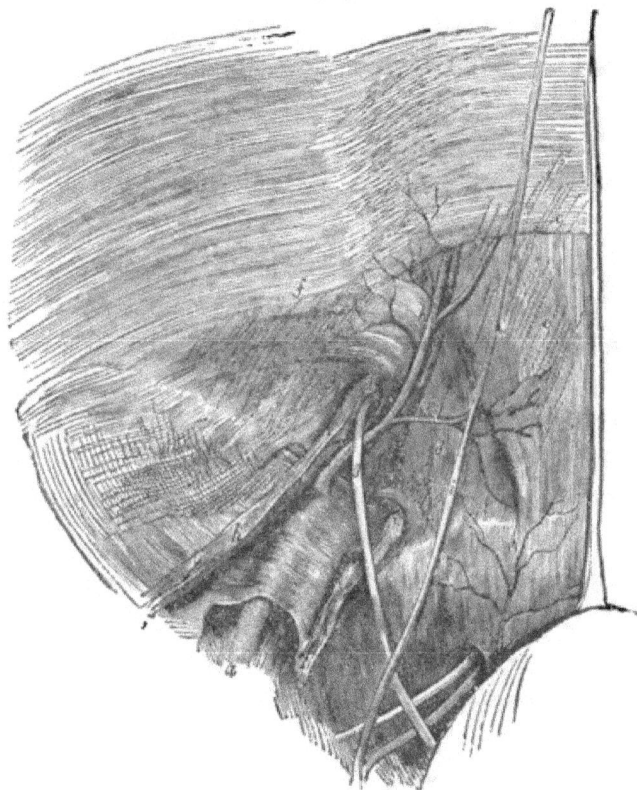

Dissection of the lower part of the abdominal wall from within, the peritoneum
having been removed. *a.* External iliac artery. *b.* Epigastric artery. *c.* Border of the posterior part of the sheath of the rectus (fold of Douglas). *d.* Conjoined tendon in the triangle of Hesselbach. *e.* Posterior surface of rectus.
f. Fascia transversalis. *g.* Vas deferens. *h.* Spermatic vessels. *i.* Obliterated
hypogastric artery. *k.* Lymphatics in crural rings. *l.* Internal abdominal ring.
(From WOOD, on Rupture.)

The position of the *internal abdominal* ring is recognized as a dimple-like depression in the peritoneum, indicating the closure of its vaginal process.   Below this

is a furrow showing the position of Poupart's ligament, and below this again a depression over the crural ring.

The cords of the obliterated hypogastric arteries are seen as ridges, passing upwards towards the umbilicus, forming the margins of the superior false ligaments of the bladder, and between them lie the remains of the urachus. On stripping off the peritoneum, the loose subperitoneal fascia is seen, in which lie the deep epigastric and circumflex ilii vessels, the latter running along the deep surface of the crural arch.

Parts concerned in Inguinal Hernia.—Inguinal hernia is described as being oblique or direct, with reference to the inguinal canal; and external or internal, with reference to its position to the deep epigastric vessels.

The *inguinal canal* is an oblique channel, about an inch and a half long in the male, and about two inches in the female, owing to the greater breadth of the pelvis, and its openings are the internal and external abdominal rings; and *the relation of the internal or deep ring to the surface* is indicated by a point taken about half or three-quarters of an inch above the centre of Poupart's ligament, along a line at right angles to it. This internal ring is an oval opening in the fascia transversalis, transmitting the cord in the male and the round ligament in the female, and is bounded above and externally by the arched fibres of the transversalis muscle, and internally by the deep epigastric vessels.

The boundaries of the inguinal canal are, in front, the integument and superficial fascia, the aponeurosis of external oblique, the internal oblique for its outer third, and a small portion of the cremaster. *Behind,* the conjoined tendon, triangular fascia, transversalis fascia, subperitoneal fat, and peritoneum. *Above,* the fibres of the internal oblique and transversalis; and *below*, Poupart's ligament and the fascia transversalis.

FIG. 27.

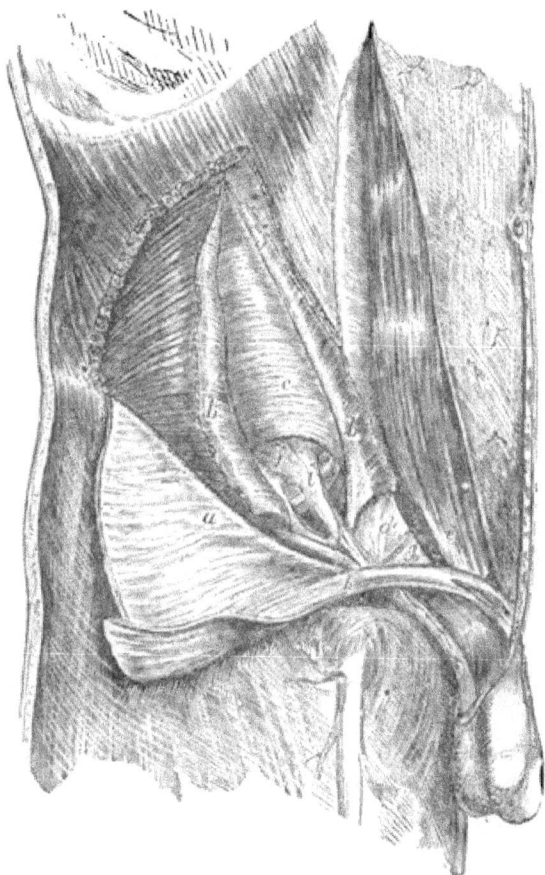

Dissection of the inguinal canal. *a.* External oblique (turned down). *b, b.* Internal oblique. *c.* Transversalis. *d.* Conjoined tendon. *e.* Rectus abdominis, with sheath opened. *f.* Fascia transversalis. *g.* Triangular fascia. *h.* Cremaster. *i.* Infundibular fascia. (From WOOD, on Rupture.)

Oblique or external inguinal hernia follows the course of the spermatic cord or round ligament, passing through both rings.

The coverings of an oblique inguinal hernia are the

same as those of the cord—viz., from without inwards:
(1.) Integument. (2.) Superficial fascia. (3.) Intercolumnar fascia. (4.) Cremaster muscle. (5.) Infundibuliform fascia. (6.) Subserous cellular tissue. (7.) Peritoneum (*sac*). If the intestine passes into the scrotum,
it is called *complete;* if retained in the canal, *incomplete*,
or *bubonocele*.

In cases of strangulation, the constriction is due to
some portion of either of the rings, or if in the canal, to
the fibres of the internal oblique or transversalis, and
any incision for the relief of the stricture should be
made *upwards*, to avoid wounding the deep epigastric
vessels or spermatic cord, which in this form of hernia
lie, the former to the inside of the neck of the sac, and
the latter directly behind it.

Fig. 28.   Fig. 29.

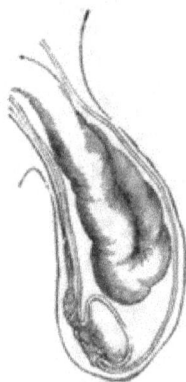 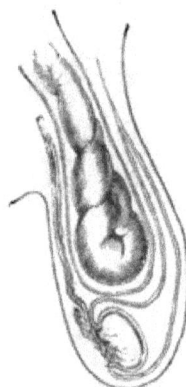

Fig. 28 —Diagram of a congenital hernia, the sac being continuous with the
tunica vaginalis testis. (HEATH.)

Fig. 29.—Diagram of an infantile hernia, showing the tunica vaginalis prolonged in front of the sac. (HEATH.)

Varieties.—Oblique inguinal hernia is liable to varieties, known as congenital, infantile, and encysted. In
the *congenital* form, the pouch of peritoneum which ac-

companies **the cord** and testis in its descent during fœtal life remains patent, and the gut falls into this pouch, and thus lies in contact with the testicle. (In congenital hydrocele the condition of the parts is the same.)

In the *infantile* form, the peritoneal pouch is only partially obliterated, and the sac descends along the inguinal canal into the scrotum, behind the pouch; hence there are three layers of peritoneum in front of the gut and its proper investment—viz., two of the tunica vaginalis testis, and the sac itself.

Direct or internal inguinal hernia differs in its course from oblique, in not passing through the inner ring, but through the space known as the triangle of Hesselbach, the boundaries of which are,—*externally*, the epigastric artery; *internally*, the outer edge of the rectus, and *inferiorly*, Poupart's ligament. This space is filled in on its inner two-thirds by the conjoined tendon, and for the rest by the fascia transversalis. Any hernial protrusion through this interval and emerging from the external ring, would have the deep epigastric artery external to its sac, and the spermatic cord internal and posterior. This form of rupture may either force its way through the conjoined tendon, or push it before it.

Coverings of Direct Inguinal Hernia.—The same as those already given in the case of the oblique variety, with the exception that the conjoined tendon takes the place of the cremaster, the infundibuliform fascia being replaced by that portion of the fascia transversalis behind or immediately contiguous to the conjoined tendon·

The seat of stricture, in strangulation, is either at the neck of the sac, at the external ring, or is due to the fissured edges of the conjoined tendon. The incision for its relief is to be made *upwards.*

16

The parts divided in an operation for strangulated inguinal hernia, would be those between the outer ring and the gut, and the sac, if so indicated by the nature of the case, and when reached, the constriction itself; although in actual practice, the condition of these structures is often so altered that this arrangement must be merely regarded as anatomical.

Structures to be Avoided.—The deep epigastric vessels and the cord.

The *position of the deep epigastric artery with regard to the abdominal parietes* is pretty much that of the superficial epigastric vessels seen beneath the integuments, and its course is indicated by a line drawn from a point a little internal to the centre of Poupart's ligament to about the middle of the space between the umbilicus and pubes.

SURGICAL ANATOMY OF THE CRURAL REGION.

In works on Descriptive Anatomy, this region is generally described and dissected as belonging to the lower limb ; but as its surgical relations essentially concern those of the contents of the abdomen, and as it has so many points in common with it, besides forming, by its deep aspect, part of the abdominal parietes, it has been thought advisable to introduce it here, and to refer back to it again when describing the superior femoral region.

The boundaries of this region are,—*above*, the crural arch; *externally*, a line passing from the anterior iliac spine to the trochanter major ; *internally*, the prominence of the adductor longus ; and *below*, a line drawn through the point of meeting of the sartorius and adductors.

These several muscular eminences inclose an irregular triangular space, sloping towards the centre, in which lie the common femoral vessels; it is the seat of crural or femoral hernia, which appears at the inner and upper part.

Dissection.—An incision skin deep is to be made along Poupart's ligament, met by one along the outer border of the adductor longus, and the flap turned down towards the sartorius. The superficial layer of superficial fascia is first met with, continuous with that over the abdomen, containing a good deal of fat, in which lie the superficial circumflexa ilii, epigastric and external pudic vessels; filaments from the external, internal, and middle cutaneous, crural and ilio-inguinal nerves, lymphatic glands, and ducts. It will be noticed that the lower chain of the lymphatic ganglia in the axis of the thigh and the glands are those which become enlarged in ulcers or injuries of the lower limb, whilst the upper series, which lie in the fold of the groin, receive the lymphatics of the genital organs, and become affected in venereal complaints. On removing this layer, the deep portion of the superficial fascia is met with attached to Poupart's ligament, and to the margins of the saphenous opening, forming a spout-like prolongation over the internal saphena vein as it lies in this opening. That portion of the superficial fascia which closes in the saphenic opening, is called the *cribriform* fascia, from its numerous perforations, due to the passage of lymphatic ducts and the superficial vessels already named. When this fascia has been removed, the *fascia lata* is next seen, a dense aponeurotic structure, attached by an outer or iliac portion to the crest and anterior spine of the ilium; and blended with the lower edge of Poupart's

ligament to the spine of the pubes and the ilio-pectineal
line, where it unites with Gimbernat's ligament. This
portion of the fascia lata forms a falciform process which
passes in front of the sheath of the vessels, and is the
outer pillar of the saphenic opening. The inner portion,
or pubic, is attached along the inferior outlet of the pelvis, and is there connected with the perineal fasciæ and
penis, and passing behind the femoral vessels, becomes
attached to the ilio-pectineal line, being thus connected
with Gimbernat's ligament, the falciform process of the
iliac portion, the fascia iliaca, and the capsule of the hip-
joint. It is through this opening that the internal saphena vein passes to join the common femoral vein.

The fascia lata attached along Poupart's ligament is
next to be detached and turned down, when the anterior
layer of the *sheath of the vessels* is brought into view,
which is a process of the fascia transversalis, and emerges
from beneath the crural arch. The posterior portion of
this sheath is formed by the fascia iliaca, and it will be
seen that it occupies the space between the psoas muscle
and Gimbernat's ligament. If the crural arch be pulled
upwards a dense band of fibres will be seen connecting
the upper layer of the sheath with the crural arch, and
reaching from the psoas to Gimbernat's ligament and
the conjoined tendon. This is the *deep* crural arch. If
now three vertical slits be made into the sheath, one
over the course of the artery externally, another over
the femoral vein, centrally, and a third a little internal
to the course of the vein, the sheath will be found to be
divided into three compartments, separated by distinct
processes. The inner compartment is termed the *crural
canal*. The interval between the femoral vein in its
compartment and the curved edge of Gimbernat's liga-

ment is the *crural ring*, the other boundaries of which
are—*above*, the deep crural arch ; *behind*, ilio-pectineal
line, origin of pectineus, and attachment of pubic portion of fascia lata. On passing the finger into the crural

FIG. 30.

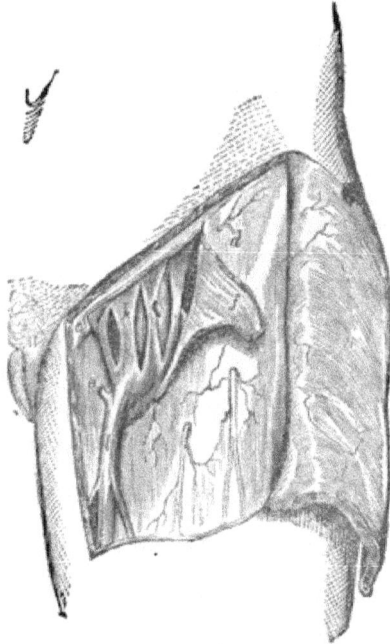

Crural sheath laid open. *a*. Middle cutaneous nerve. *c*. Placed to the inner
side of Gimbernat's ligament. *d*. Iliac portion of fascia lata. *e*. Pubic portion of
fascia lata. *f*. Margin of saphenous opening (turned back). *k*. Femoral sheath
opened by three incisions. *l*. Saphena vein. (From WOOD, on Rupture.)

ring, the inferior portion of Gimbernat's ligament can
be felt along the ilio-pectineal line, considerably behind
the ring.   The crural ring is blocked in above by a thin
fascia derived from the subperitoneal, termed the *septum
crurale*, which transmits ducts of glands.   The crural
canal is usually occupied by a lymphatic gland.

In the employment of *taxis* for the reduction of a

crural hernia, it must be remembered that the direction
of the crural canal is *downwards*, and slightly *forwards*
and *outwards;* moreover, that, in order to relax the orifices of this canal, the thigh must be flexed upon the
pelvis, adducted and rotated inwards.

The course taken by a crural hernia is as follows:
First, passing into the femoral ring, it descends vertically in the femoral canal as far as the saphenic opening;
next, being here prevented from passing further along

Fig. 31.

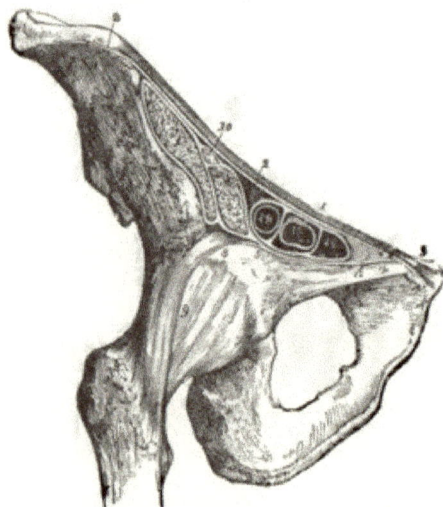

Section of the structures which pass beneath the femoral arch. 1. Poupart's
ligament. 2, 2. Iliac portion of the fascia lata, attached along the margin of the
crest of the ilium, and along Poupart's ligament as far as the spine of the pubes
(3). 4. Pubic portion of the fascia lata, continuous at 3 with the iliac portion,
and passing outwards behind the sheath of the femoral vessels to its outer border
at 5, where it divides into two layers; one is continuous with the sheath of the
psoas (6) and iliacus (7); the other (8) is lost upon the capsule of the hip-joint
(9). 10. The anterior crural nerve. 11. Gimbernat's ligament. 12. The femoral
ring within the femoral sheath. 13. Femoral vein. 14. Femoral artery; the
two vessels and the ring are surrounded by the femoral sheath. (From WILSON.)

the sheath of the vessels, it is directed forwards and
subsequently upwards, upon the external pillar of the
opening and Poupart's ligament.

The coverings of a crural hernia are—sac, subserous areolar tissue, septum crurale, sheath of vessels, cribriform fascia, superficial fascia, and integuments.

The seat of stricture may either be the sac itself, or the junction of the falciform process with Gimbernat's ligament, or the outer margin of the opening; and in dividing the obstruction, the incision is to be made *upwards* and *inwards*.

An irregular course of the *obturator artery* bears a very important relation to the crural ring, should it arise by a common trunk with the deep epigastric, and when it courses along the border of Gimbernat's ligament, in order to gain the thyroid foramen. In this case the neck of the hernia would be surrounded by an arterial ring, and in an operation for the relief of stran-

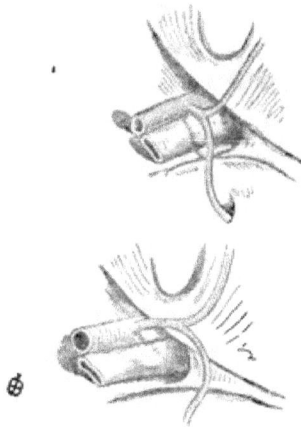

Irregular course of obturator artery.

gulation, might run great risk of being wounded. Practically, however, it would probably recede, unless the knife were roughly pushed past the posterior aspect of the ring.

Besides herniæ, the fold of the groin is the seat of other *tumors*, of which an accurate knowledge of the anatomy of the region affords the chief help towards the diagnosis; such as aneurism, the pointing of a psoas abscess, an inflamed lymphatic gland, cysts, enlargement of the bursæ beneath the tendon of the psoas, which is often connected with the hip-joint.

Artificial Anus.—In cases of strangulated hernia, whether inguinal or crural, when the bowel has become

FIG. 33.

Sketch of artificial anus.
1. Mesentery. 2. Éperon. 3. Opening of artificial anus.

gangrenous, or if the intestine has been wounded, and given way in a state of gangrene, surgical principles indicate the removal of this portion, and the subsequent formation of an artificial anus, and the following condition generally results, which explains the impediments in the way of its healing : The edges of the gut become attached to the aperture in the abdominal wall, and that generally at an angle which soon becomes acute. As the upper portion of the intestine only transmits fæces, the lower remains as a useless tube, and consequently contracts in its calibre. That portion of the bowel to which the mesentery is attached, becomes drawn out into a spur-like process (*éperon*); which acting as a valve, serves to direct the fæces out of the body, but not on into their proper channel.

SURGICAL ANATOMY OF THE ILIAC FOSSA (EXTRA-PERITONEAL).

The iliac fossa forms the inner aspect of the region just described, and is of great surgical interest. From within outwards the structures successively met with are, the peritoneum, the subperitoneal fat and cellular tissue, which contains a large amount of fat, allowing the peritoneum to be readily stripped off, in ligature of the vessels, the common and external iliac arteries and their veins, the fascia iliaca, which is a continuation of the fascia transversalis, and has attachments to the crest of the ilium, Poupart's ligament, and vertebral column. Beneath the fascia are the circumflex ilii artery and veins, the iliacus and psoas muscles, in the substance of which is the anterior crural nerve; upon the former muscle are seen the external cutaneous and the ilio-inguinal, and upon the latter the genito-crural nerves, and the nutrient vessels derived from the ilio-lumbar.

The chief points of surgical importance refer to ligature of the vessels, the common and external iliac arteries, and abscesses.

The Common Iliac.—The course taken by this vessel, with regard to the surface of the body, is from a point to the left of the umbilicus, and in a line with the iliac crests and the centre of Poupart's ligament. After the intestines and peritoneum have been raised, it will be seen that the aorta bifurcates, or does generally, on the left side of the fourth lumbar vertebra, consequently the vessels of the right and left side differ somewhat in length, the right being slightly the longer, and lies rather more obliquely across the body of the fifth lumbar vertebra. Their length is about two inches, more or less, and passing downwards and outwards, at the sacro-iliac synchondrosis divide into external and internal iliacs; the vena cava inferior lying to the right side, and being formed by the union of the two common iliac veins, the right common iliac artery crosses their junction, rendering the relation of the vessels on the right side more intimate, the vein projecting externally to the artery above, and being internal to it below, whilst on the left side the vein lies below and internal to its artery. Both are crossed at their bifurcation by the ureter.

The relations of the common iliac of the *right* side: In *front* of it, the peritoneum, ilium, sympathetic plexus, and ureter; *externally*, the cava, right common iliac vein, psoas magnus; *behind*, junction of common iliac veins, obturator nerve. On the *left* side it has, in *front*, peritoneum, sympathetic, rectum, inferior mesenteric artery, ureter; *externally*, psoas magnus; *internally*, left common iliac vein; *behind*, left common iliac vein, obturator nerve.

Ligature of Common Iliac Artery.—In order to reach this vessel a curved incision is recommended, commencing from just above the middle of Poupart's ligament to a point an inch or so above and to the inner side of the anterior superior spine, or, commencing from a point close to the anterior superior spine, towards the edge of the rectus. The first incision divides the integuments; next the external oblique, the internal oblique, and the transversalis are to be divided in succession and to an equal extent. Beneath this latter is the transversalis fascia, which is to be carefully pinched up, nicked, and a director insinuated between it and the peritoneum. It is then to be divided to the length of the incision in the muscles (the deep circumflex iliac artery will stand a chance of being cut). The peritoneal bag and its contents are to be then pulled away towards the middle line, when the vessel will be seen just beyond the sacro-iliac synchondrosis. The needle is to be passed *from within outwards*.

The collateral circulation would be maintained by the inosculation of the lateral sacral and middle sacral, epigastric and internal mammary, aortic intercostals and lumbar, the ilio-lumbar and last lumbar, the obturator, with its fellow of the opposite side, and the epigastric, gluteal, and sacral, hæmorrhoidal of the internal iliac, with the superior hæmorrhoidal of the inferior mesenteric and the vesical of the opposite sides, and the uterine and ovarian in the female.

The External Iliac Artery.—The course of this vessel would be indicated on the surface of the body by a line extending from either side of the umbilicus to the centre of Poupart's ligament. It commences at the bifurcation of the common iliac, and extends to the crural arch, where it becomes common femoral.

Its relations are,—in *front*, intestines and peritoneum, a considerable quantity of loose areolar tissue, the spermatic vessels, the genito-crural nerve (near Poupart's ligament), the circumflexa ilii vein, and a chain of lymphatics; *externally*, the psoas muscle, and fascia iliaca; *internally*, the external iliac vein, vas deferens, and lymphatics; *behind*, the external iliac vein and psoas magnus muscle.

Ligature of External Iliac.—An incision (curved for

preference) is to be made, commencing at the middle of
Poupart's ligament, and at about an inch above it, to a
point just beyond the anterior superior spine of the
ilium.   The structures divided and the method of dividing them are those described in ligature of the common
iliac, with this exception, that here they are more aponeurotic, from being nearer Poupart's ligament; and the
deep epigastric artery is in this instance in danger of
division, if the incision be made too near the rectus
muscle.

When the peritoneum and intestines have been pulled
away from the point where the ligature is to be applied,
it often happens that neither artery nor vein is to be
found; in this case they have followed these structures,
and will be discovered lying adherent to the under surface of the peritoneal bag, with the ureter, from which
they must be cautiously separated.   The needle is to be
applied *from within outwards*.

Collateral Circulation after Ligature of External Iliac:
The gluteal anastomoses, with the external circumflex
from the profunda femoris; the ilio-lumbar, with the
circumflexa ilii; the obturator, with the internal circumflex from the profunda; the ischiatic, with the perforating and circumflex branches of the profunda; the internal pudic, with the superficial and deep external pudic,
and the internal circumflex from the profunda; and the
deep epigastric with the superior epigastric from the internal mammary.

SURGICAL ANATOMY OF THE LUMBAR REGION.

The lumbar region forms the posterior portion of the
abdominal parietes, and is of surgical importance from
the application to its anterior aspect of the abdominal

viscera, and from the numerous fasciæ which enter into its formation, and the relation of these fasciæ to abscess, &c.

The limits or boundaries of the region may be defined as follows : *above*, the lower border of the last rib; *below*, the crest of the ilium ; *externally*, a line drawn through the end of the first rib perpendicularly to the iliac crest; and *internally*, the line of the spinous processes of the vertebræ.

Dissection.—On making a dissection of the region indicated by these limits, from the integument towards the abdominal cavity and its contents, the following structures would be met with : The integument, tough and thick ; the subcutaneous cellular tissue, containing a great deal of fat, excepting along the middle line ; the aponeurotic origin of the latissimus dorsi and serratus posticus inferior, part of the external oblique, and common aponeurotic attachment of the internal oblique and transversalis, the mass of the erector spinæ muscles, a considerable number of vessels and nerves lying between these muscles, the attachment to the transverse processes of the lumbar vertebræ of the middle lamina of the aponeurosis of the transversalis muscle, the quadratus lumborum muscle and ilio-lumbar ligament, branches of the lower part of the dorsal and upper part of lumbar plexuses, psoas muscles, transversalis fascia, a large quantity of fat and cellular tissue separating the kidney from the parietes, the kidney itself, with the ureter and spermatic or ovarian vessels, and in the front of the psoas, the colon, the ascending on the right, and the descending on the left. The ascending colon is generally inclosed in peritoneum, which forms, by its attachment to the spine, a mesocolon, whilst the descending is covered only

anteriorly and laterally, and is for that reason selected
for the **operation of opening the colon in** the left loin
(Amussat).

That portion of the spinal column which is inclosed
between these regions, has upon its anterior surface the
crura of the diaphragm, covered by peritoneum and subperitoneal fat and fascia, the vena cava ascendens, abdominal aorta, and on either side the chain of the sympathetic, the thoracic duct, the receptaculum chyli, vena
azygos major, and a large number of lymphatic ganglia.

Lumbar Colotomy.—The colon may be reached either
by a transverse incision (Amussat), or by a longitudinal
one (Hilton, Callisen).

Structures divided in Amussat's **Operation.**—A point is
taken midway between the crest of the **ilium** and the
last rib at the outer edge of the erector spinæ, varying
in length according to the development of the individual,
and is directed outwards, at first dividing the integuments, the aponeurotic origin of the latissimus dorsi, and
some few fibres of the external oblique; next the origin
of the internal oblique and transversalis, and a portion
of the quadratus lumborum and its fascial investment.
After the transversalis fascia has been divided, a quantity
of loose cellular tissue and fat is seen, which, being
scratched through, exposes the colon. During life, however, the distension is generally so great that the bowel
bulges into the wound.

In the vertical incision, which is made about four
inches or so external to the spinous processes of the vertebræ, the structures divided would be, the integument,
the aponeurotic origin of the latissimus dorsi, the origin
of the internal oblique, tendon, transversalis and trans-

versalis fascia. The kidney may be mistaken for the contents of the bowel.

Lumbar Fascia.—The peculiar arrangement of the fasciæ in the lumbar region is of great importance surgically, from the control it has over the course taken by the pus in lumbar abscess. This fascia is the posterior aponeurotic portion of the transversalis, and posteriorly gives attachment to the internal and external oblique

FIG. 34.

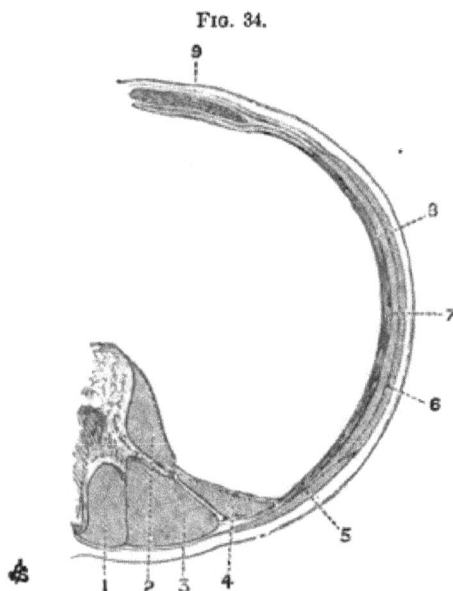

Arrangement of lumbar aponeurosis at level of third lumbar vertebra. 1. Sacro-lumbalis. 2. Psoas magnus. 3. Longissimus dorsi. 4. Quadratus lumborum. 5. Latissimus dorsi. 6. External oblique. 7. Internal oblique. 8. Transversalis. 9. Rectus.

and latissimus dorsi muscles. From its inner edge, two laminæ, derived from it, are attached to the transverse processes of the lumbar vertebræ, and inclose the quadratus lumborum, the posterior lamina separating it from the erector spinæ, whilst the posterior portion of the erector spinæ is covered in by the aponeurosis of the latissimus dorsi.

In spinal *caries* of the lumbar region, the pus, by perforating the quadratus lumborum, between the last rib and crest of ilium, passes backwards, and being bound down by the various fascial laminæ just mentioned, forms a tumor, usually flat, broad, and slightly elevated ; occasionally the pus finds its way between the abdominal muscles and points *above* Poupart's ligament.

Although the psoas muscles essentially belong to the lumbar region, yet psoas abscesses do not necessarily *arise* in them, the course taken by the pus being in a great measure governed by the fascial investment they obtain there ; this investment is that of the fascia iliaca, which, as regards the muscles, is attached above to the ligamenta arcuata interna, internally to the sacrum, being continued over the muscles to the crural arches, beneath which the pus gravitates, and ultimately points *external* to the femoral vessels.   Sometimes it passes through the sacro-sciatic notch to the nates.

The relation of the nerves to the spinal column, emerging as they do either through the intervertebral spaces or sacral foramina, readily accounts for the course taken by the pus in these abscesses.

Iliac abscess is a collection of matter either in the cellular tissue, between the iliac fascia and peritoneum, or between the fascia iliaca and iliacus muscle, and points above Poupart's ligament, near the anterior superior spinous process of the ilium.

SURGICAL ANATOMY OF THE PELVIS.

Those portions of the true pelvis and its contents which come within the province of the surgeon are more particularly, the genito-urinary apparatus, the termination of the bowel and its inferior boundary, the perineal, and the anal regions. Those considerations of its anatomy which more particularly concern the accoucheur, may be advantageously referred to special works upon the subject, such as its various measurements, &c.

Its walls present surgically two surfaces for examination, on either of which operative proceedings may be instituted—(1) An *external*, partly free, consisting inferiorly of the perineum, laterally of a region belonging partially to the lower limb, and posteriorly, of the sacral and coccygeal region ; and (2) an *internal*, consisting of the peritoneal surface of the perineum, and soft parts lining its bony structure, which include the pelvic viscera and the great vessels.

The external genitals may be regarded as appendages to the pelvis, and will be conveniently treated of before entering upon its immediate anatomy, and such parts of its lateral boundary as are evidently common to it and to the lower limb will be included in the description of the latter.

EXTERIOR OF PELVIS—PELVI-PERINEAL REGION. SURGICAL ANATOMY OF THE SCROTUM.

Structure.—The various tissues entering into the structure of the scrotum are met with on dissection in the following order: The *skin*, very thin, lax, and rugose ; the *superficial fascia*, with which it is closely associated, continuous with that covering the abdomen and perineum, and is in this region entirely destitute of fat. The *dartos*, consisting of loose areolar tissue, in which is a considerable amount of unstriped muscular fibre, very vascular, is continuous with the superficial fascia of the crural region and perineum, and sends a septum inwards which divides the scrotal bag into two cavities, thus separating the testes. The dartos is connected with the subjacent parts by delicate areolar tissue, allowing of the free movement of the scrotal structures over the testes, and owing to the corrugating power it has over the scrotal tissues, it is difficult to approximate the edges of incisions when made in its structure. A *fibro-cellular* coat

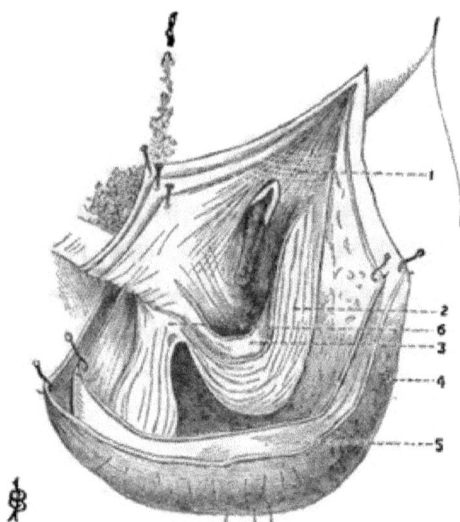

Fig. 35.

Tissues forming the scrotum. 1. Fibres of external oblique. 2. Dartos. 3. Fibrous tunic. 4. Skin. 5. Superficial fascia. 6. Dartos, forming septum between the testes.

has been described, which is the continuation of the external spermatic fascia, but it is very thin, and not easily followed beyond the cord. Beneath this layer is the *cremasteric fascia*, derived from the lower border of the internal oblique and gubernaculum during the descent of the testis, and the *fascia propria* or *infundibuliform*, a derivative of the fascia transversalis. All these structures are interunited by a very lax cellular tissue, which not only allows of their free movement over each other, but over the cord and testis.

The scrotal tissues are not very sensitive, and have not much vitality; consequently in erysipelatous inflammation or urinary extravasation they rapidly become gangrenous. When the urethra gives way from unrelieved retention, or from ulceration of its walls, the urine is driven by the sudden contraction of the bladder into the cellular interval between the scrotal and abdominal fasciæ, and its direction is limited by the attachments of the fascia already named. Commencing at first in the scrotum, it ascends over the pubes and abdomen and cellular tissue of penis, but cannot descend down the thighs, owing to the attachment of the deep layer of superficial fascia along Poupart's ligament. In the case of wound of the urethra from without, such, for instance, as a blow in the perineum, not only is the urethral tube itself ruptured, but the fasciæ enveloping it, often to an unlimited extent; hence the urine may follow almost any course, and not restrict itself to anatomical relations.

The course taken by urine is often rather theoretical than real. For instance, in actual practice the inflammation set up by the escape of urine, whether owing to a false passage, or to the lesion of the urethral walls, causes rapid perforation from gangrene, resulting in loss of substance. Moreover, there is always a considerable number of natural openings and passages, undescribed by the anatomist, but readily found out by an infiltrating fluid, through which it passes and frequently shows itself in the most unexpected places. (*Vide* Perineum.)

The *spermatic cord* consists of external spermatic fascia, cremaster, infundibuliform fascia, vessels, lymphatics, nerves, and the vas deferens, which can be readily isolated from the other structures, being recognized by its whipcord-like texture. The *arteries* of the cord are the spermatic, deferential, and the cremasteric; of these the spermatic supplies the substance of the testis, passing into it either through the tunica albuginea, or through the back of the mediastinum testis. The *veins*, passing from the testis, unite in forming the *pampiniform* plexus, which passes into a single trunk forming in the body of the cord, and terminating, the right one in the vena cava, and the left in the left renal vein.

The veins of the cord are very liable to a varicose condition (*varicocele*), which is due to several anatomical causes : (1) their tortuous arrangement and free anastomoses at their emergence from the gland; (2) their want of support, lying as they do in the loose areolar tissue, which allows of the weight of the contained column of blood obliterating their valves; (3) the pressure they receive in their passage through the inguinal canal. It is a matter of fact that the left spermatic veins are more liable to this condition than the right, the causes assigned being, that the left testicle hangs lower, and that its upward current of blood meets at a right angle that proceeding from the kidney in the left renal vein, and that they are liable to constant pressure from the contents of the sigmoid flexure of the colon.

The *lymphatics* are numerous and large, and terminate in the lumbar glands, which become rapidly affected in malignant disease of the testis.

In addition to the coverings mentioned, the testicle has that derived from the peritoneum; the tunica vaginalis, the result of the descent of that organ into the scrotum. That portion of the peritoneum which is thus cut off from the bag, remains as a shut sac, investing the outer surface of the testis (*tunica vag. testis*), and is reflected on to the scrotum (*tunica vag. reflexa*). Collections of fluid (*hydrocele, hæmatocele*) are thus related to the testis—namely, that they are above and in front of it, unless there be any abnormality, such as an inversion of the testis. In removal of the testis (*castration*), the retraction of the cord into the abdominal cavity, owing to the action of the cremaster muscle, is liable to give trouble, unless it be firmly secured.

The *structures divided in the operation of castration* would be as follows: (1) the scrotal tissues, with the vascular and nervous supply—viz., the superficial perineal vessels and nerves, inferior pudendal nerve, superficial external pudic vessels; (2) the structures entering into the formation of the cord.

SURGICAL ANATOMY OF THE MALE PERINEUM.

This region is to be studied when the body is placed in what is known as the lithotomy position—that is, with the legs flexed on the thighs, and the thighs on the pelvis, in order that the parts to be examined be on the stretch.

In the erect position the superficial aspect of the region becomes a mere fold.

There is considerable difference in the descriptions given by authors as to what the limits of the perineum really are: some including all those structures which close in the inferior outlet of the pelvis, its entire floor in fact; others dividing this lozenge-shaped space into

two triangles, by a line passing from one tuberosity of
the ischium to the other, in front of the anus, and call-
ing all in front of it the perineum, and all behind it, the
anal, or ischio-rectal region. It is proposed to adopt
the latter method in this description, and it will be seen
that, although it may appear arbitrary as far as the superficial layer of fascia is concerned, as the dissection
proceeds deeper towards the inner aspect of the pelvis,
the arrangement is a natural one, as adapting itself to
the special contents of each triangular space. The
boundaries of the anterior portion, which is convex in
the middle, owing to the position of the bulb of the
urethra, are—below, a line passing horizontally in front
of the anus from one tuberosity of the ischium to the
other; and laterally, the pubic rami, meeting at the
symphysis; thus forming a triangular interspace, which
is itself divided by the median raphé into two equal
parts, the left being the one in which the lateral operation of lithotomy is performed, supposing the operator
to be right-handed. This raphé indicates the course
taken by the urethra, and is a most valuable landmark
in all operations about the perineum. The sides of this
triangular space are about three inches and a half long,
and about three inches across the base, and a line drawn
from the apex of the triangle to its base, about three
inches.

Superficial Dissection of the Perineum.—A staff having
been introduced into the bladder, and the hair shaved
off the nates, an incision is to be made through the integument in the middle line, a second at right angles to
it in front of the anus, and a third, parallel to the latter
across the base of the scrotum. The flaps thus formed
are to be reflected outwards, when the subcutaneous cel-

lular tissue, which is very adherent to the integument
along the middle line, is met with. The superficial
fascia may be conveniently divided into two layers: the
upper containing a good deal of fat (which considerably
augments the depth of the perineum in some cases); is
continuous with that of the scrotum and thighs, and
in it, or immediately beneath it, lie the superficial perineal vessels and nerves, whilst the deeper layer has important attachments and is more membranous in texture.
Externally, it is attached to the rami of the pubes and
ischium, outside the crura and erectores penis; behind,
it is continuous with the deep perineal fascia, or triangular ligament of the perineum, after turning round the
transversalis perinei; and in front it is continuous with
the dartos and fasciæ at the root of the penis; a septum
derived from it passes inwards, dividing the posterior
part of the space beneath this layer of fascia into two,
but is, however, ill-defined in front. This cellulo-fatty
layer is a favorite seat of abscess from urinary or other
infiltration.

The attachments of this fascia are very important as
directing the course of the urine in extravasation, from
rupture of the urethra anterior to the triangular ligament,
into the cellular tissue of the scrotum and penis.

Along the mesial line this fascia is intimately attached
to the bulb of the urethra, and in cases where the bulb
has been injured, the urine, after extravasation, readily
finds its way along the spongy portion of the urethra to
the glans, with which it is continuous. On reflecting
this layer of fascia, in the middle of the space, are the
acceleratores urinæ muscles, enveloping the bulb, and on
either side are the erectores penis, passing from the inner
aspect of the ascending ramus, and covering the lower

portions of the crura. Lying somewhat obliquely to the central tendon are the transverse muscles of the perineum. This central tendon is a white fibrous knot, and acts as a *point d'appui* for the acceleratores urinæ and external sphincter muscles; it is situated in the median line between the urethra and the anus. Lying between and upon the erectores penis and the acceleratores urinæ, are the trunks of the superficial perineal vessels and nerve, and some inosculating branches of the inferior pudendal, and on or below the posterior border of the transversus perinei muscle lie the transverse perineal vessels and nerve. The acceleratores urinæ are separated from the deep layer of superficial fascia by a thin aponeurotic layer. This muscle, which compresses the bulb, and empties the bulbous urethra, is generally described as consisting of three portions, commencing from the median raphé,—an anterior, which passes round the penis to be inserted on its upper aspect, sending an expansion which compresses the dorsal vein; a middle, which incloses the inferior portion of the urethra, and passes between it and the body of the penis; and a posterior, which is attached to the anterior surface of the triangular ligament.

It will be observed that the perineal muscles of one side form a triangular space, having the triangular ligament as its floor, while from its outer angle emerge the superficial perineal vessels, and the transverse perinei vessels and nerves coming to the surface; the relations of the space are of importance, as in the lateral operation of lithotomy the first incision traverses it.

If the accelerator urinæ muscle be now carefully detached by making an incision along the raphé, the bulb of the urethra is exposed, small in childhood, and large in advanced age, and closely bordering on the margin of the anus; this fact is of importance, as there is greater danger of wounding the bulb in the lateral operation of lithotomy in old persons. The crus penis and its muscle should now be drawn outwards (or entirely removed), and the rectum drawn downwards, when the anterior surface of the *triangular ligament* will be seen as a tough bluish-white structure—the fibres of which are nearly all transverse—allowing of the structures between the two layers being readily seen through it in favorable subjects; its base is directed towards the rectum, it is attached in the middle line to the central tendon of the perineum, and laterally to the rami of the ischium and pubes, having a free margin on either side of the central tendon, which is continuous with the deep layer of superficial fascia; its apex is directed upwards, and is connected with the periosteum in front of the symphysis pubis. It is perforated at about an inch below the symphysis by the urethra, with which it is intimately connected, and which here changes its direction, and between the urethra and the symphysis lie the vessels of the penis, the dorsal vein or veins in the centre, on either side of it the dorsal arteries, and most externally are the dorsal nerves.

The bulb may now be separated from the triangular ligament (if both sides of the perineum be available), turned upwards, and the triangular ligament itself carefully detached from the bone. A considerable plexus of veins is usually first met with, and care must be taken not to divide it, as the blood would obscure the view of *the parts between the layers of the triangular ligament,* which consist of the following structures,—a plane of muscular fibres, variously described by different authors, as the levator and compressor urethræ, surrounding the

18

membranous portion of the urethra, which lies between
these layers of fascia, and receives a prolongation from
each ; the deep transversus perinei ; the internal pudic
artery and nerve, the former giving off the artery to the
bulb and to Cowper's gland ; the artery to the corpus
cavernosum, and the dorsal artery of the penis ; and just
below the urethra, Cowper's glands, their ducts, and the
subpubic ligament. Beneath this layer of muscular
fibres and vessels is the posterior layer of the triangular
ligament, derived from the pelvic fascia.

Note.—Before proceeding with the deep dissection of
the perineum it will be found expedient to study the
anal region.

Dissection of the Anal or Ischio-rectal Region.—An
incision is to be made commencing just in front of the
anus, round which it is to be carried to the tip of the
coccyx, another at right angles to it, immediately behind
the termination of the rectum, and the flaps reflected
outwards. As the external sphincter is incorporated
with the integument, great care must be taken not to
remove it, on dissecting off the flap ; this is next to impossible, however, if the dissection be conducted in the
usual manner, as the skin of the whole of the perineum
posteriorly is interwoven with muscular fibres.

In the middle line of this space lies the anus, the mucous membrane of which is not seen, in the normal state,
during life. If, however, the anus be gently opened at
the junction of the mucous membrane with the skin, a
pale line is to be seen, marking the position of the internal sphincter, a thickening of the muscular fibres of
the lower portion of the rectum. This band of muscular fibres plays an important part in preventing the healing of ulcers and fissures of the rectum, by keeping the

tissues on the stretch. These ulcers are generally situated about a quarter of an inch or so from the verge, just within the sphincter, and generally either in front of, or at the side of the coccyx. The treatment consists in dividing the fibres with a view to relieve the tension. In operating on the female, care must be taken in making the incision, if on the anterior wall of the bowel, on account of the proximity of the vagina. If the verge of the anus be carefully examined it will be seen to contain a large number of follicles, suppuration in which is often mistaken for fistula—a disease which is frequently difficult to diagnose correctly without a good view of the interior of the gut, of which an inch or an inch and a half must be exposed for the purpose; and it is a fact of great surgical importance that the internal opening of a fistula is always within this distance of the orifice.

The external opening of the rectum is occasionally wanting (atresia ani).

The lower, or perineal portion of the rectum, is not much more than an inch in length; it curves back below the prostate, and is uncovered by peritoneum. Just above the anus the rectum is considerably dilated, a condition increased by age and constipation. The folds of mucous membrane, in the empty state of the rectum, so overlap, that considerable difficulty may be experienced in passing the finger or a bougie through them; one fold in particular often obstructs the finger, at about an inch and a half above the aperture. The mucous membrane of the lower end of the bowel is very loose, and readily admits of the burrowing of matter. It must be borne in mind that the curve of the bowel above mentioned necessitates caution in the introduction of an enema-tube or other instrument, which should be directed obliquely *from below upwards* **and** *forwards,* and afterwards, *upwards* **and** *backwards.*

A dilated condition **of the** inferior hæmorrhoidal **veins at the lower part of the anus** constitutes *external piles.*

The superficial fascia, tough and strong, and containing a great deal of fat, has cutaneous vessels and nerves passing through it. The external sphincter muscle is seen attached posteriorly to the tip of the coccyx, and inclosing the margin of the anus, is inserted into the **central tendon** before mentioned. This circular band of fibres is about one **inch in** breadth. Between the bowel and the tuberosities of the **ischium is** the *ischio-rectal fossa,* which contains a quantity of loose fat and cellular tissue, and lying across it, and passing to the margin of the anus, are the superficial hæmorrhoidal vessels and nerves, which are liable to give a good deal of **trouble from** hemorrhage, when **cut** in operations for fistula, &c. On cleaning out the space its shape and boundaries can **be defined.** In shape it is somewhat triangular, about **an inch in** width, and about two inches deep; inferiorly **its base is** formed by **the** integuments of the region, and **its apex,** directed upwards, corresponds to the interval between the lower border of the obturator internus, covered by the obturator fascia, and the outer surface of the levator ani, covered by the anal fascia. Its boundaries are,—externally, the tuberosity of the ischium and obturator fascia; internally, the sphincter ani, levator ani, covered by anal fascia, and coccygeus; anteriorly, the triangular ligament; and posteriorly, the gluteus maximus and great sacro-sciatic ligament.

Lying in a fold derived from the obturator fascia, on the outer wall of the fossa, the trunks of the internal pudic vessels and nerve can be easily felt, grooving the inner aspect of the tuberosity of the ischium.

In the external incision for lateral lithotomy the knife sinks into the ischio-rectal fossa, and will divide the superficial hæmorrhoidal vessels and nerves. Abscesses have a great partiality for the ischio-rectal fossæ, and often burrow to a most remarkable extent. *A fistula in ano*, is the sinus left on the contraction of the cavity of such an abscess. True fistulæ exist external to the sphincter, and always extend as far up as its upper border. They are called complete or incomplete according as their openings are situated; thus, in the former case, one opening is in the rectum and the other on the surface of the body, generally near the anus; in the latter, there is an opening into the bowel and none external, or the converse.

The operation for its cure consists in passing a knife through the fistulous track into the bowel, and cutting through all the tissues between the edge of the knife and the interior of the gut. These tissues are—the pseudo-mucous membrane of the fistula, the external sphincter, some few fibres of the levator ani, the branches of the inferior hæmorrhoidal vessels and nerves, the internal sphincter, and the mucous membrane of the inner bowel and its vessels.

Deep Dissection of the Perineum.—Those structures before described as lying beneath the anterior layer of the triangular ligament may be now removed, when its posterior layer will come into view. This is derived from the pelvic fascia, and covers the hinder part of the membranous urethra, and outer surface of the prostate gland; it is attached below -to the anterior layer, forming a pouch on either side of and below the urethra, in which lie Cowper's glands.

On detaching this posterior layer, the anterior fibres of the levator ani are seen, passing by the sides of the prostate, and uniting on its perineal surface with the muscle of the opposite side, and blending at the central tendon with the fibres of the external sphincter and transverse perineal muscles. The central fibres are inserted into the side of the rectum, interlacing with the sphincters, and the posterior are attached to the coccyx and median raphé behind the rectum. These muscles and the triangular ligament shut in the inferior outlet of the pelvis.

In order to obtain a view of the relations of the structures which lie at the inferior outlet of the pelvis, as they would be met with in a surgical operation, the rectum should be detached from its connections by dividing the anterior and lateral portions of the levator ani, and pulled backwards, when the under surface of the prostate, the neck and base of the bladder, vesiculæ seminales, and vasa deferentia will be seen (Fig. 36).

The general form of the normal *prostate* is that of a chestnut, with its base directed towards the bladder, and its apex towards the symphysis, having its longest diameters antero-posteriorly, and at its base transversely. Its inferior surface rests flat on the triangular ligament and membranous portion of the urethra; its upper surface, slightly concave, is intimately connected with the bladder and ejaculatory ducts, which lie together in the middle line immediately behind it. Its anterior surface corresponds to the deep layer of the triangular ligament (pubo-prostatic ligament); the posterior surface is separated from the rectum simply by a little cellular tissue and is applied to the neck of the bladder; and its sides are in relation with the levator ani and with the pelvic fascia. The prostate is invested by a fibrous capsule, derived from the pelvic fascia. The density of this capsule accounts for the intense pain of prostatic abscesses, and forces the pus to find its way, unless relieved, into the urethra. In opening these abscesses in the perineum, there is a possibility of urinary fistula, and in the event of their bursting in the perineum, such fistulæ are certain to form.

The *position of the prostate gland* is readily determined by passing the finger into the rectum, and if healthy is generally felt about as far up as the second joint of the forefinger reaches, whilst in some forms of enlargement, the upper border of the gland will be far out of reach of the entire finger.

The *relations of the bladder and the rectum*, within the reach of the finger, are of great importance; thus, in cases of retention of the urine, when it is necessary to perform the operation of *puncture per rectum*, the distended bladder is felt overlapping the posterior margin of the prostate at a point where, if the puncture be made in the mesial line, no injury to surrounding parts could take place, as the instrument would pass between the vesiculæ seminales, and perforate a space (*trigone vesicæ*) where these structures are neither covered by pelvic fascia nor by peritoneum. The digital examination of the bladder *per rectum* assists the surgeon in the detection or dislodgment of vesical calculi, in guiding the point of a catheter or sound in cases of difficulty, and in the detection of prostatic abscesses, or growths.

Parts concerned in Lateral Lithotomy in the Adult.— The object to be attained, is that of opening the bladder at one particular spot, its neck, and for the reason that if opened at any other, urinary infiltration into the areolar tissue of the pelvis will take place. The incisions then must be made in the most direct way, to allow of (1), the position of the staff, which has been introduced into the bladder, being felt ; (2), the neck of the bladder being opened, and room obtained for the extraction of the stone.

The perineum having been shaved, the skin and integuments are to be steadied and rendered tense with the fingers of the left hand, and the point of the knife is to be entered about an inch and three-quarters in front of the anus, a little to the left of the middle line, and carried through the skin, in a direction downwards and outwards, midway between the anus and the tuberosity of the ischium. The left forefinger is next to be pushed into this external wound, with the double purpose of feeling for the position of the groove of the staff in the urethra, and for the purpose of pushing the rectum inwards and backwards out of the way. When the groove is recognized, the knife, lying flat under the introduced finger, is pushed into the urethra just in front of the prostate, and when the point is felt to be in the groove, it is made to slide along it towards the bladder, dividing in its course, the membranous urethra and left lobe of the prostate to the extent of an inch. The forefinger is now to be pushed along the groove, through the edges of the deep wound, and insinuated into the incision through the prostate ; the staff is then withdrawn by the assistant in charge of it, whilst the finger passes into the cavity of the bladder. The forceps are next guided by the under surface of the finger into the bladder. When the stone is felt, the blades must be opened, the finger gradually withdrawn, and an attempt made to catch it in its long axis if possible. When caught, it is to be slowly and firmly drawn out, without hurry, with a slight to and fro motion in a direction *downwards*,

Fig. 36.

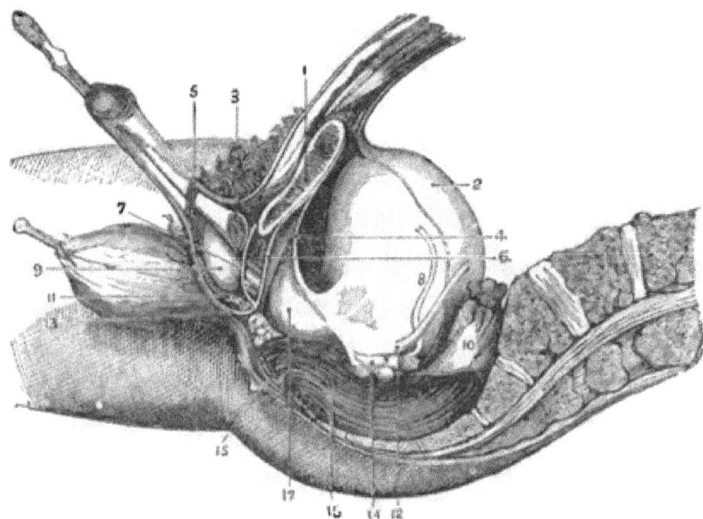

S ction of pelvis to the left of the median line at the pubes, and through the middle line at the sacrum. 1. Section of left pubic bone. 2. Peritoneum on bladder. 3. Left crus penis (cut). 4. Pelvic fascia, forming anterior ligaments of bladder. 5. Part of accelerator urinæ. 6. Posterior layer of triangular ligament, or pelvic fascia forming the capsule of the prostate. 7. Anterior layer of triangular ligament, or deep perineal fascia. Between 6 and 7 are seen the following: Membranous urethra, deep muscles of urethra (insertion), and Cowper's gland of the left side. 8. Vas deferens. 9. Bulb of urethra. 10. Rectum. 11. Cut edges of accelerator urinæ and transversus perinei. 12. Left ureter. 13. Reflection of deep layer of superficial fascia round transversus perinei. 14. Left vesicula seminalis. 15. Cut edge of levator ani. 16. Rectum. 17. Prostate. (HEATH.)

towards the floor, and not horizontally towards the pubic arch. After the operation, a searcher or sound is to be introduced to find out whether the bladder is free.

Structures Divided in the Lateral Operation of Lith-otomy.—The superficial incision divides the skin and

superficial fascia, **inferior** hæmorrhoidal vessels and
nerves, which lie superficially in the anterior part of the
ischio-rectal fossa, transversus perinei muscles and vessels, and superficial perineal vessels and nerves. The
structures divided *on* the staff are the lower part of the
triangular ligament, deep transversus urethræ muscle;
the *deep* part of the incision divides the membranous
urethra, the substance of the prostate, and vessels
around it.

Structures to be Avoided.—The *bulb,* or the *rectum,*
which stands a risk of being wounded, if the first incision be too near the middle line; the internal pudic
artery, if on the other hand the deep incision be made
too far externally; the *artery* to the bulb, if it be made
too far forward; and the entire breadth of *prostate,* with
its *capsule,* and the *ejaculatory* ducts, if it be carried too
far backwards or downwards.

If the entire breadth of the gland and its capsule were
divided, the urine would be infiltrated beneath the peritoneum.

The depth of the perineum between the neck of the
bladder and the integument varies from rather more
than an inch to four inches, and between the tuberosities
of the ischium from two inches or less, to four.

In the child, the pelvis being narrow, the perineum is
narrow also; and the neck of the bladder comparatively
high up, whilst the peritoneum descends very low between the bladder and the rectum. The bladder itself is
more conical in shape, and is rather an abdominal than
a pelvic viscus, and its connections with the surrounding
parts are very loose. Hence the difficulty experienced
in getting into the bladder in lateral lithotomy in children, and the danger of pushing the prostate before the

finger and tearing it from the membranous portion of
the urethra. There is also danger of cutting the urethra
considerably anterior to the point indicated as the exact

FIG. 37.

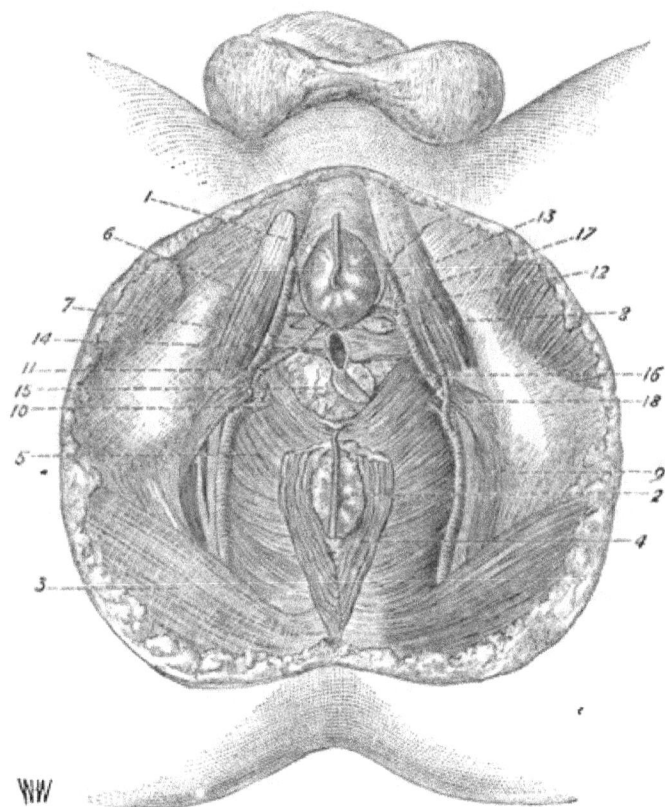

(The bulb is slightly raised and the rectum drawn backwards, in order to make
clear the membranous urethra and prostate, which are shown incised as in the
lateral operation of lithotomy.) 1. Bulb. 2. Rectum. 3. Gluteus maximus. 4.
External sphincter. 5. Levator ani, its anterior fibres raised to show the prostate. 6. Erector penis. 7. Wilson's muscle. 8. Cowper's gland. 9. Trunk of
internal pudic. 10. Superficial perineal artery. 11. Artery to bulb (abnormal).
12. Artery to bulb. 13. Continuation of internal pudic artery. 14. Membranous
urethra divided as in the lateral incision. 15. Prostate gland, with its plexus of
vessels. 16. Incision in the prostate gland as in the lateral operation. 17. Anterior layer of triangular ligament. 18. Transversus perinei muscle. (RICHET.)

position for entering the bladder, therefore always, in
children, the external incision should be made as large
as possible, that the relative position of the parts be
clearly made out.

SURGICAL ANATOMY OF THE PENIS AND MALE URETHRA.

The integument of the penis consists externally, of
very lax, loose skin, destitute of fat, which at the corona
glandis is reflected over the glans, and has an internal
mucous surface continuous with that of the glans; immediately below the meatus urinarius this membrane is
gathered into a fold, the *frœnum preputii*. Behind the
corona, and in the sulcus, are a number of glands which
secrete the smegma. Beneath the skin is a layer of loose
muscular fibres, analogous to the dartos, arranged circu-
larly and lying in loose cellular tissue. Beneath this is
a tough, elastic fascia, enveloping the entire body of the
organ, sending in a process beneath the urethra and
corpora cavernosa, continuous with the superficial fascia
of the perineum, incorporated at its root with the suspensory ligament; between the two laminæ of which lie the
dorsal vessels and nerves. The upper portion of the
body of the penis is composed of the corpora cavernosa,
which, arising from the inner aspect of the horizontal
rami of the pubes, unite along the mesial line, this union
being marked by a septum, called the *septum pectiniforme*, which, however, is wanting in front. The corpora cavernosa terminate in the front, by a rounded margin, which projects into the base of the glans. The
inferior portion of the body of the glans is formed by the
corpus spongiosum, containing the urethra. It com-

mences in front of the triangular ligament at the bulb, and lying between and below the united crura, terminates at the glans.

The *arteries* of the penis are the dorsal, which lie in the dorsal furrow, supplying the integument, and afterwards pierce its fibrous investment, near the corona ; the arteries to the corpora cavernosa, and the arteries to the bulb. The *veins* are very numerous, and are superficial and deep, the former passing into the dorsal vein, which lies between the two dorsal arteries, and generally terminates in the internal saphena; and the deep, after piercing the triangular ligament, terminate in the prostatic plexus.

The *lymphatics*, with which the organ is richly furnished, accompany the dorsal vessels and pass into the ganglia of the fold of the groin.

The *nerves* lie external to the arteries on the dorsum, and are freely distributed to the body and glans.

The penis is often the seat of an arrest of development, one form of which, where the anterior wall of the urethra is wanting, is termed *epispadias ;* and where the superior wall is wanting, and generally associated with extroversion of the bladder, *hypospadias.* Occasionally the prepuce completely incloses the glans, excepting a minute orifice through which the urine passes (*congenital phimosis*).

SURGICAL ANATOMY OF THE MALE URETHRA.

Supposing in the first instance the parts removed from the body, for the sake of examining the canal, the urethra may be described as extending from the neck of the bladder to the meatus urinarius, and is from eight to nine inches in length, and for general division consists of one portion belonging to the penis, and of another belonging to the perineum; with the former is included the spongy portions, and with the latter the membranous and prostatic.

Spongy Portion.—Commencing from the orifice of the urethra, a vertical slit provided with two lip-like margins, the urethral tube is seen at its most constricted portion. On examining the floor of the canal within the

FIG. 38.

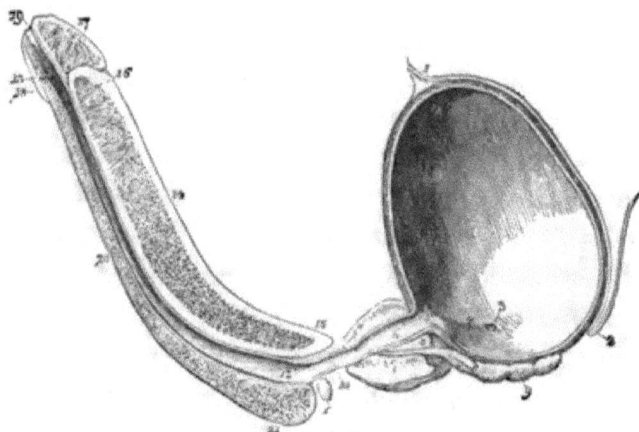

Longitudinal section of the bladder, prostate gland, and penis. 1. Urachus. 2. Recto-vesical fold of peritoneum. 3. Opening of the right ureter. 4. A slight ridge formed by the muscle of the ureter. 5. The neck of the bladder. 6. Prostatic portion of the urethra. 7. Prostate gland. 8. The common ejaculatory duct. 9. Right vesicula seminalis; the vas deferens is cut short. 10. Membranous portion of the urethra. Its direction is the reverse of this when *in situ.* 11. Cowper's gland of the right side, with its duct. 12. Bulbous portion of the urethra. 13. Fossa navicularis. 14. Corpus cavernosum. 15. Right crus penis. 16. A portion of the septum pectiniforme. 17. The glans penis. 18. Corona glandis. 19. Meatus urinarius. 20. Corpus spongiosum. 21. Bulb of the corpus spongiosum. (WILSON.)

meatus, a considerable dilatation is found, termed the *fossa navicularis,* and on the roof of this part of the urethra is the orifice of a large mucous pouch, the *lacuna magna.* Behind this dilatation the canal averages about a quarter of an inch in diameter, and is throughout studded, particularly on its floor, with the orifices of glands (glands of *Littré*), opening forwards. About five inches behind the orifice is another pouch-like dilatation contained within the bulb, into the floor of which open the ducts of Cowper's glands.

Membranous Portion.—This is the narrowest portion of the tube throughout its length, excepting the sphincter-like orifice, and is contained between the layers of the triangular ligament; it measures about three-fourths of an inch along its upper, and half an inch along its lower surface, and consists of mucous membrane, elastic, erectile, and muscular tissue.

The prostatic portion is the widest and most dilatable portion of the urethra; it is about an inch and a quarter long, and lies nearer the upper than the lower portions of the gland, and its tube is of greater calibre in the middle than at either entrance or exit; on its floor, at the neck of the bladder, is the uvula vesicæ, in front of which is a ridge of mucous membrane, rather deeper behind than before, called the *veru montanum* or *caput gallinaginis*, having on either side of it a pouch or sinus, into which open the prostatic ducts. At the fore part of the veru montanum is a cul-de-sac, running upwards and backwards beneath the middle lobe, containing on its floor the openings of the ejaculatory ducts; it is called the *sinus pocularis*.

Next, let the urethra be examined as existing during life, as it would present itself to the surgeon.

The urethra may be thus divided into a penile and a perineal portion, and the individual lying on his back, the usual position for catheterism, its direction can be conveniently described as an ascending portion, ter-

minating at the root of the penis and descending at the
bulb; and a descending, comprising the membranous and
prostatic. Thus the points where the urethra changes direction are at the root of the penis and bulb, and it is in
this portion of the canal that false passages are most frequently made. These curves disappear on catheterism;
the first by merely raising the penis, and the second on the
depression of the handle of the instrument between the
thighs. So resilient are the urethral walls that a perfectly
straight instrument can be readily introduced into the
bladder. The urethral canal is distant from the under
border of the symphysis about half an inch or a little
more, and is consequently about half an inch or a little
more below the vesical aperture, which corresponds with
the lower border of the symphysis, and is about an inch
and a quarter behind it. When not in use the walls of
the urethra touch each other, excepting at the orifice of
the meatus urinarius and in the bulb, where they are
separated by a narrow interspace.

Catheterism of the Male Urethra.—If the urethra be
healthy, the sound or catheter will pass by its own
weight, and require scarcely any urging. The instrument
is to be introduced into the orifice of the urethra, and
pressed gently onwards until it has traversed the canal
for four or five inches, when the handle is to be brought
to the middle line close to the abdomen, in order that
the point may traverse the curve below the symphysis;
the handle is then to be brought gently down towards
the surgeon, when it should glide into the bladder.
The great point is to keep the extremity of the instrument traversing the *upper* wall of the urethra. Besides,
being less movable, experience shows that in cases of
stricture, the upper wall is less liable to be affected than

the base and sides. In introducing a small instrument, the position of the lacunæ must be borne in mind, particularly that of the lacuna magna, on the upper wall of the navicular fossa, as it is liable to intercept its point; and if force be employed, it might pass *beneath* the mucous coat. In cases of difficulty, by passing the finger into the rectum the point of the instrument can be directed into the bladder, on account of the close relation of the membranous portion of the urethra and the rectum, and the readiness with which the catheter, or sound, can be felt through it.

Catheterism in the female is a very easy proceeding generally, and the little papillar orifice which is situated about an inch below the clitoris, in the back part of the vestibule, being detected, a straight instrument is readily slipped in without exposing the patient. When any difficulty is experienced, it is owing to some deviation of the canal or of the neck of the bladder, caused by some tumor pressing upon the parts, which are very mobile. Occasionally fecal accumulations in the rectum have been known to prevent micturition, from pressing upon the neck of the bladder, so that in such cases of retention, when there is difficulty or impossiblity of introducing the female catheter, careful vaginal examination must be instituted.

In *lithotomy in the female* the vesico-vaginal operation is the best, provided the resulting fistula be properly treated. There is, however, a chance of permanent incontinence of urine, on account of the function of the urethral sphincter being destroyed by the incision into the neck of the bladder having been made too freely. It must be borne in mind that when the bladder is fully contracted, the septum between the bladder and vagina is very

19

limited; moreover, in this condition the openings of the ureters are brought very low down, and might be implicated in the incision, which is on no account to be transverse. The length of the female urethra is about an inch and a half, curving slightly below the symphysis, with its concavity upwards, and having an average diameter of about a quarter of an inch, and being highly distensible, very frequently calculi can be extracted through it. There is a good deal of difficulty experienced generally in using a lithotrite in the female bladder, owing to the fact of its muscular coat being so thick and strong and its urethral sphincter so weak, that the urine or water injected for the purpose of operation, escapes past the instrument, leaving no cavity; moreover, the bladder forms a fossa on both sides of the neck of the uterus.

The arrangement of the *perineal* aponeuroses in the female is as follows: The superficial layer of superficial fascia is continuous with that of the nates, thigh, and abdomen; whilst the deeper layer is firmly attached to Poupart's ligament, the ischio-pubic rami, and to the lower border of the perineal septum. These fasciæ cover in the labia majora, which are very analogous to the scrotum in the male, and being attached above to the external abdominal ring, herniæ pass in them, known as pudendal or labial herniæ. This deeper layer of fascia is continuous over the ischio-rectal fossæ. Where the two layers of superficial fasciæ unite with the lower borders of the perineal septum to form the perineal body, they are joined by the ischio-perineal ligament, and it forms a support, or *point d'appui*, for the perineal muscles.

Abscesses in the female perineum are of two kinds,— diffuse, in the superficial perineal fascia, which readily spread in all directions; and circumscribed abscess of the vulvo-vaginal gland, which would be seen as an oval projection on the side of the vestibule.

The sacro-coccygeal region offers for surgical consideration a common arrest of development of the neural arches of the sacrum, constituting *spina bifida;* and the nerves (cauda equina) are in this region usually connected with the sac. Hence, if the operation of puncture be deemed advisable, it should always be made on one side of the sac, and at its lowest part.

CAVITY OF PELVIS.

The pelvic cavity contains those viscera, the inferior relations and apertures of which have been described in the preceding section—viz., the bladder, rectum, and vagina, with its appendages, the superior surface of the perineum forming its inferior boundary. The soft parts lining its bony walls, the obturatores interni, pyriformes, and levatores ani muscles, are invested with the reflexion of the pelvic fascia, upon which lie the peritoneum and subperitoneal cellular tissue, the arrangements of which are of considerable importance surgically.[1]

The *pelvic fascia,* which is continuous with the fascia iliaca, is itself a continuation of the transversalis fascia. It is attached to the brim of the true pelvis, and round the margin of the obturator internus muscle. At a curved line between the spine of the ischium and the pubes, this fascia splits to inclose part of the origin of the levator ani muscle, the external lamina of which (obturator fascia) is applied to the inner surface of the

---

[1] The surgical anatomy of the uterus and ovaries will be better studied in special works on obstetrics.

obturator internus muscle, and passing beneath the obturator vessels and nerve, completes the obturator canal; it
is attached below to the pubic rami, where it forms an
investment for the internal pudic vessels and nerve. It
afterwards sends a thin fascia, the anal, over the lower

FIG. 39.

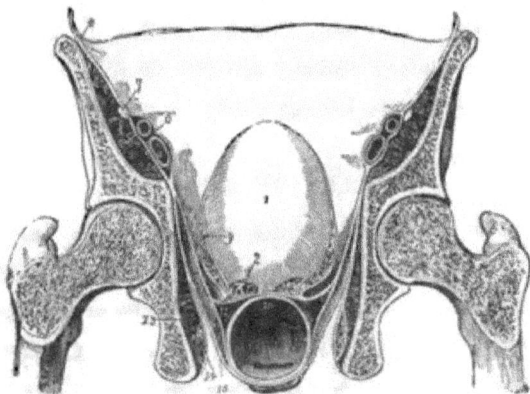

Transverse section of the pelvis seen from behind, showing the distribution
of the pelvic fascia. 1. Bladder. 2. Vesicula seminalis of one side divided. 3.
Rectum. 4. Iliac fascia, covering in the iliacus and psoas (5), and forming a
sheath for the external iliac vessels (6). 7. Anterior crural nerve excluded from
the sheath. 8. Pelvic fascia, splitting into the recto-vesical and obturator layers.
9. Recto-vesical layer, forming the lateral ligament of the bladder of one side,
and a sheath to the vesical plexus of veins. 10. A layer of fascia passing between
the bladder and rectum. 11. A layer passing around the rectum. 12. Levator
ani. 13. Obturator internus, covered in by the obturator fascia, which also forms
a sheath for the internal pudic vessels and nerve (14). 15. Anal fascia, investing
the under surface of the levator ani. Figures 14, 15, are placed in the ischio-
rectal fossa. (WILSON.)

surface of the levator ani, which is to be seen covering
it in the ischio-rectal fossa. The internal lamina (the
recto-vesical), is continued over the upper surface of the
levator ani, over the bladder and sides of prostate, and
lower end of rectum. The pubo-prostatic ligaments are
formed by two short rounded bands extending from the
capsule of the prostate to the posterior aspect of the

symphysis; the capsule of the prostate is formed from the lateral attachment to it of this fascia, which also incloses the vesico-prostatic plexus of veins. It is this portion of the pelvic fascia which it is so important to avoid dividing posteriorly in the operation of lateral lithotomy, for by so doing the urine would find its way into the loose areolar tissue between the rectum and the bladder. The presence of the prostatic plexus of veins is often a serious source of danger. Posteriorly, the pelvic fascia is continuous over the pyriformis muscle and sacral plexus, being perforated by the internal iliac artery and vein.

The inlet of the pelvis is somewhat heart-shaped, well padded along its upper border by the psoas and iliacus muscles, while posteriorly, in the middle line, is the promontory of the sacrum, or sacro-vertebral angle, which can be readily felt through the abdominal parietes in thin persons. Between the bladder and rectum is the *recto-vesical pouch*, formed by the pelvic fascia, which corresponds posteriorly where it is broad to the interval between the iliac arteries; it is narrow in front between the rectum and the bladder, and extends as far as the vesiculæ seminales, and, in front and behind, to the tip of the coccyx. Its relation to the orifice of the anus is important, and must be referred to the condition of the bladder, which, if distended, will raise the pouch further into the pelvic cavity than its usual level, which is about four inches above the anal aperture. Some coils of the ileum and sigmoid flexure of the colon fill in the space. The peritoneum affording no investment to the lower end of the rectum, the neck, base, and anterior surface of the bladder, or the front and inferior portion of the posterior wall of the vagina, permits of operative pro-

ceedings upon these viscera, without danger of wounding
it.   The space beneath the membrane varies considerably
in different parts of the floor of the pelvis: thus, in front
and at the sides it is tolerably closely applied to the
underlying pelvic fascia, leaving, however, a considerable
interspace in front of the bladder, the point selected for
puncturing that viscus above the pubis.   Behind, and
above the anal region, there is a considerable interval,
containing a great deal of loose cellular tissue, which
allows of the distension of the rectum, and of the internal iliac vessels and their branches, the ureters, sacral,
sympathetic and hypogastric plexuses, and the origin of
the pyriformis muscle.

SURGICAL ANATOMY OF THE BLADDER.

The bladder is situated in the mesial line, beneath the
pelvic fascia and peritoneum, and lies obliquely from
above downwards.   Being attached to the pelvic floor
by its body and base only, it is freely movable, but the
urachus and anterior reflexion of the peritoneum limit
its mobility posteriorly.   When empty, the bladder lies
deep in the pelvis as a flattened sac, with its apex reaching up to the symphysis pubis, but when distended, its
relations are considerably altered, and are of great surgical importance.   When moderately full, it is round and
partially fills the true pelvis, but when greatly distended
it rises up into the abdomen, perhaps even reaching to
the umbilicus, and becomes curved forwards.

In *children*, as has been before stated, it is rather an
abdominal than a pelvic viscus, and is conical in shape,
owing to the position of the but recently obliterated
urachus.

In front of the bladder, between it and the pubes, is
a quantity of lax cellular tissue, the reflexion of the
peritoneum from its anterior surface, and the non-attach-
ment of this membrane to it allows of its dilatation, and
is of great practical importance in percussing the bladder. It is here that puncture over the pubes and the
" high " operation of lithotomy are practiced, but the
frequency of urinary infiltration is greatly against the
latter proceeding. Its posterior surface is entirely covered by peritoneum, and corresponds in the male to the
rectum, and in the female to the vagina, and is in relation
with some convolutions of the small intestine, which lie
in the recto-vesical pouch. Laterally, the peritoneum is
applied to the bladder above and behind the crossing of
the obliterated umbilical artery, around which it is reflected. The vasa deferentia pass along the side, cross
the obliterated umbilical artery, and lie to the inner side
of the ureter.

The base of the bladder lies upon the anterior surface
of the rectum, from which it is separated merely by a
thin layer of cellular tissue in the middle, and laterally
by the vesiculæ seminales and vasa deferentia, the former
of which are intimately adherent to it, and form two
sides of a triangle, the base of which is directed upwards
and backwards, and its apex towards the prostate. It
is at this spot that puncture of the bladder by the rectum is performed. The neck of the bladder is surrounded by the prostate, and is directed obliquely forwards and downwards.

The cavity of the bladder presents at its base the *trigone vesicæ*, an equilateral triangle formed by the oblique
openings of the ureters posteriorly, and by the urethra
anteriorly. It is through this triangular space that the

trocar enters in puncture per rectum. The trigone is perfectly smooth, and free from rugæ.

Immediately behind the trigone, is the deepest part of the bladder, the *bas fond*—not very much marked in children, but forming a considerable pouch in old persons, in which the urine settles, causing considerable irritation. It is in this pouch that calculi lodge generally.

In the female the bladder is rather larger than in the male. It has no *bas fond.* The neck is lower, and it and the posterior portion of the bladder lie on the vagina. Fistulous openings occasionally occur between the bladder and vagina, or rectum and vagina.

Internal Iliac Artery.—The artery of the region is the internal iliac, which furnishes, with the exception of the middle sacral, all the vascular supply of the walls, the soft parts, and viscera. It is given off from the common iliac, between the sacro-iliac synchondrosis and the sacro-vertebral angle.

After birth the vessel consists of two trunks, an anterior and a posterior—a subdivision which takes place opposite the great sacro-sciatic notch. The branches given off from the *anterior* are those to the bladder and prostate; superior vesical (the pervious portion of the fœtal hypogastric artery), middle and inferior vesical, the middle hæmorrhoidal to the rectum, the obturator, the internal pudic and ischiatic, and the uterine and vaginal in the female. Those given off from the posterior trunk are the gluteal, ilio-lumbar, and lateral sacral. The gluteal, ischiatic, and internal pudic leave the pelvis by the great sacro-sciatic notch, passing between the sacral plexus of nerves.

Relations of the Internal Iliac Artery.—The internal

iliac has, in front, the peritoneum and ureter (rectum on left side); *externally*, the psoas muscle, and obturator nerve; *behind*, the internal iliac vein, lumbo-sacral nerve, and pyriformis muscle; on the right side the vein is more external.

Ligature of Internal Iliac Artery.—This vessel is reached by the same incision as that for the common iliac, and the bifurcation being found, the space is very limited upon which the ligature can be placed; the short thick trunk passes downwards and backwards as far as the upper border of the great sacro-sciatic notch.

The *external* iliac vein lies just in the bifurcation of the common into external and internal iliac, and is liable to be in the way of the needle in passing the ligature. The lumbar and sacral arteries chiefly carry on the circulation after its ligature.

20

CHAPTER VI.

SURGICAL ANATOMY OF THE LOWER EXTREMITY.

THE region of the hip includes, anteriorly, the superior femoral region, or the upper third of the front of the thigh, immediately below Poupart's ligament; posteriorly, the gluteal; and internally, the ischio-pubic or obturator. These several regions cover in the articulation of the hip-joint. The region of the hip-joint may be described as having for its limits, the iliac crest and sacral groove, Poupart's ligament, and below, a line encircling the limb, just below the fold of the nates.

Surface Markings.—*Anteriorly,* are the muscular prominences formed by the tensor vaginæ femoris externally; *internally* the adductors and gracilis, crossed obliquely below by the sartorius. *Posteriorly,* is the swell of the gluteus maximus; and *internally,* the tuberosity of the ischium, and the rami of the pubes and ischium, covered by their muscles.

The position of the *trochanter major* with regard to the several bony projections of the region should be carefully studied in every position of the limb. Its situation is marked by a deep depression, when the individual is standing upright with the heels together, and its differences of relation in flexion, extension, adduction, and abduction, should be compared. These relations are obviously of the utmost importance in the diagnosis of dislocation or fracture connected with the hip-joint.

If the exact relations of the great trochanter with the several osseous prominences observable on the pelvis in a normal state be examined, it will be noticed that if the femur be flexed at a right angle, and at the same time slightly adducted, the apex of the great trochanter corresponds with a line drawn from the anterior superior iliac spine to the tuberosity of the ischium, and that this line divides the cotyloid cavity (which, with respect to the surface, may be regarded as occupying the central position between the anterior superior spine of the ilium, the spine of the pubis, and the tuberosity of the ischium) into two equal parts. This line corresponding to the centre of the cavity, will serve as a guide to an appreciation of the extent of displacement in dislocation. Thus supposing the head of the femur be placed behind the cotyloid cavity, this line, instead of corresponding with the apex of the trochanter major, would correspond with a point nearer its base. The extent of the displacement, then, will be measured by the prominence of the great trochanter behind this line.

SURGICAL ANATOMY OF THE SUPERIOR FEMORAL REGION.

The superficial dissection of this region has been already described in the chapter on inguinal and crural herniæ, as it was considered convenient to associate it with the region of the abdomen (*vide* Abdomen); but the deeper surgical relations are those of the common femoral and upper portion of the superficial femoral vessels, or *Scarpa's* space.

Scarpa's Space or Triangle.—The dissection of the parts of femoral hernia having been completed, and the fascia

lata detached, a triangular space is exposed which has
for its limits the following : its *base*, the crural arch ; its
external boundary, the sartorius ; its *internal*, the adductor longus. In the centre of this triangle so formed,
passing from the middle of its base to the apex (the
meeting of the sartorius and adductor longus), lie the
femoral vessels. The floor upon which they rest is formed
from without inwards by the iliacus, psoas, pectineus,
adductor longus, and part of the adductor brevis muscles. The anterior crural nerve lies in the furrow between the iliacus and psoas muscles. The femoral artery
lies external to and a little superficial to its vein. The
sheath of the femoral vessels, which has been before described (*vide* Crural Hernia), ceases at the division of the
common femoral into superficial and deep, and is formed
anteriorly by a prolongation of the fascia transversalis,
and posteriorly by the fascia iliaca, which furnish septa
between the common femoral vessels. The inner margin
of the psoas separates the artery from the hip-joint, and
passing behind the sheath is the branch of the anterior
crural nerve to the pectineus. Lying on the pectineus,
and outer surface of the adductor longus, are the deep
external pudic vessels, branches of the common femoral,
beneath the pubic portion of the fascia lata.

On the outer side of the sartorius is the tensor vaginæ
femoris, passing obliquely outwards, and backwards, to
be inserted into the fascia lata ; and between it and the
sartorius is the upper portion of the rectus femoris, with
some branches of the external circumflex vessels passing
outwards into its substance. Passing inwards towards
the articulation will be met with, from without, inwards,
beneath the sartorius and vessels, the lower portion of
the combined tendons of the psoas and iliacus in their

sheath ; a cellular interspace between them and the outer
border of the pectineus, in which lie the internal circumflex vessels, the pectineus, and the adductor brevis, also
separated by a slight interval. Beneath, the upper portion of the rectus and the external circumflex vessels,
the upper part of the vasti, the neck of the femur, and
the anterior portion of the articulation. Immediately
behind the psoas and iliacus tendon and the pectineus,
are large bursæ separating them from the joint. Beneath
the pectineus and adductor brevis are the obturator vessels and nerve, the obturator externus, and portion of
the adductor magnus muscles.

Femoral Artery in Scarpa's Space.—The vessel is a
continuation of the external iliac, and enters the space
below Poupart's ligament, at a point midway between
the anterior superior spine of the ilium and the symphysis pubis, and it lies in the crural sheath for about an
inch and a half or two inches, which sheath separates it
from the fascia lata and inguinal glands.

The common femoral vein lies to its inner side above,
but gets behind it lower down. The anterior crural
nerve lies about half an inch external to it. It lies at
first on the psoas and afterwards on the pectineus, but
separated from it by the femoral vein and profunda vessels. It usually gives off four superficial branches—
the superficial epigastric, circumflex iliac, and the superficial and deep external pudic. The *profunda* or *deep*
femoral generally arises from the outer and back part of
the common trunk, about an inch and a half or two
inches below the crural arch ; at its commencement the
vessel is on the outside of the femoral vessels ; but it
soon passes behind, and finally reaching the inside,
courses downwards and backwards among the adductor

FIG. 40.

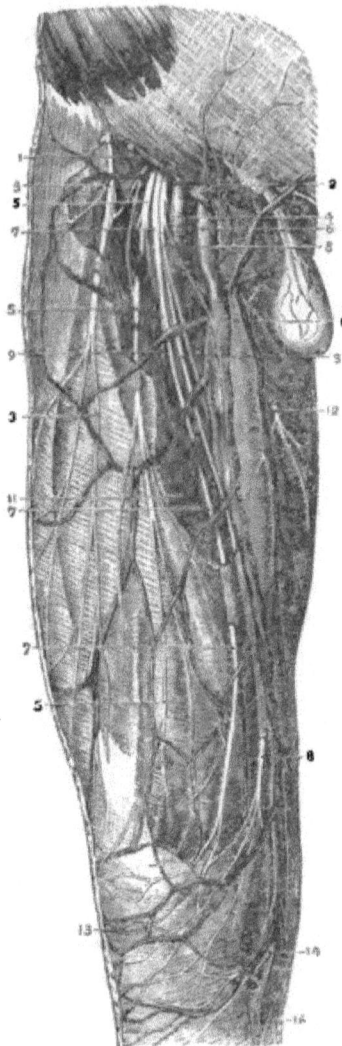

muscles. It rests on the iliacus, pectineus, and adductor brevis, and passing between the adductor longus and magnus, terminates in a small twig that pierces the magnus. The profunda gives off, the *external circumflex*, which arising from its outer side, passes outwards between the branches of distribution of the anterior crural nerve, below the sartorius and rectus, and divides into three series of branches,—ascending, descending, and transverse. *The internal circumflex* is given off from the inner and back part of the profunda, passes between the pectineus and psoas muscles, and opposite the tendon of the obturator externus, it gives off two branches: one an ascending, inosculating with the obtu-

Superficial dissection of the front of the thigh. 1. Poupart's ligament. 2. Superficial branches of femoral artery. 3. External cutaneous nerve. 4. Femoral artery. 5, 5, 5. Middle cutaneous nerve. 6. Femoral vein. 7, 7, 7. Outer division of internal cutaneous nerve. 8, 8, 8. Inner division of ditto. 9. Branch to sartorius. 10. Saphena vein. 11. Sartorius plexus. 12. Cutaneous branch of obturator nerve. 13. Patellæ. 14. Patellar branch of long saphena. 16. Long or internal saphenous nerve. (HEATH.)

rator, and a descending, muscular; and the vessel itself passes into the gluteal region between the quadratus femoris and adductor magnus, inosculating with the ischiatic, external circumflex, and superior perforating vessels. The *perforating* arteries: The first is generally given off from the profunda, just above the tendon of the adductor brevis, between it and the pectineus, and pierces the adductor magnus; the second pierces the adductor brevis and magnus; and the third is given off below the adductor brevis, and pierces the adductor magnus. The inosculations of these vessels will be considered in the description of the thigh and buttock.

Compression of the femoral artery in the upper third is easily effected, either just as it passes over the pubes, where the pressure should be made obliquely backwards, on account of the surface of the bone being inclined slightly forwards, or just below Poupart's ligament, at a point where it is very superficial, being separated from the acetabulum and neck of femur by the psoas in extension of the thigh.

Ligature of the Femoral Artery in Scarpa's Space.— Except in the case of a wound, the *common* femoral is rarely tied, owing to the number of small superficial and muscular branches, affording but little chance of a good coagulum being made. For popliteal aneurism, &c., the superficial femoral is tied just at the point before it passes beneath the sartorius. (By the superficial femoral is meant that portion of the vessel between the giving off of the profunda and the point where it becomes popliteal.)

The knee being slightly bent, an incision of about two or three inches in length is to be made over the course of the artery, dividing the skin, superficial fascia, and fat. Next the fascia lata is to be divided, when the oblique fibres at the edge of the sartorius will be seen, and which are to be drawn aside in order to give room. Some branches of the anterior crural nerve are generally spread out over the course of the vessel, and occasionally the internal saphena nerve crosses it at this point. The sheath is next to be opened, only so much so as to allow of the easy passage of the aneurism needle round the vessel, and to avoid wounding more of the *vasa vasorum* than is absolutely necessary, thus interfering with the nutrition of its coats. The needle should be passed *from within outwards.*

Collateral Circulation after Ligature of the Femoral Artery in Scarpa's Space.—The external circumflex from the profunda anastomoses with the gluteal and circumflex iliac, the internal circumflex with the obturator ischiatic, and superior perforating, and the vessels in the popliteal space with the comes nervi ischiadici.

The *tumors* in Scarpa's space, which might be mistaken for aneurisms, are enlarged glands, cysts, psoas abscess, enlargement of the bursa below the psoas, and hernia. Femoral herniæ lie to the inner and upper side of the vessels; psoas abscesses point external to them. Inguinal herniæ may be mistaken for crural, owing to the circumstance that adhesions taking place from any cause between the aponeuroses may divert their course; it is far more common for crural herniæ to resemble inguinal (*vide* Crural Hernia).

SURGICAL ANATOMY OF THE GLUTEAL REGION.

This region is of great surgical importance from its intimate relations with the hip-joint, and the control the muscles have over its several movements; it has for its limits, in *front*, the **anterior** superior spine of the ilium, and the margin of the tensor vaginæ femoris; *superiorly* the crest; *posteriorly* the posterior superior spine of the ilium, the sacrum, and the middle line of the body; *below* the tuberosity of the ischium, and fold of the nates.

Dissection.—On removing the skin, there is a considerable amount of fat, particularly over the tuberosity of the ischium, in which lie a number of cutaneous nerves, supplying the integument; the fascia lata, which is thin over the gluteus maximus, and very thick in front over the gluteus medius, to which it gives origin; next, the gluteus maximus, and the anterior and superior portion of the medius.

Parts beneath the Gluteus Maximus.—Beneath the *gluteus maximus* lie, first of all, a fibro-cellular layer, continuous with the subperitoneal cellular tissue through the great sacro-sciatic notch, a portion of the gluteus medius, sacro-sciatic ligaments, pyriformis muscle, with the sciatic vessels and nerves emerging below its inferior border; the obturator internus, with its satellite muscles, the gemelli; the internal pudic vessels and nerve with the nerve to the obturator internus, the quadratus femoris, the tuberosity of the ischium with the origin of the hamstring muscles, the great trochanter covered by a large bursa mucosa, which separates it from the gluteus maximus, and part of the origin of the vastus internus, the superficial gluteal vessels, and the anastomoses of the external circumflex with the gluteal vessels, the gluteus minimus muscle, the posterior part of the articulation of the hip-joint; and beneath the quadratus femoris, the obturator externus and the anastomosing branch of the internal circumflex.

The gluteal artery, generally the larger terminal branch

of the posterior division of the internal iliac, passes out
of the pelvis, at the upper part of the great sacro-sciatic
notch, and lies between the gluteus minimus and pyriformis.   It may be the seat of aneurism, either idio-
pathic or traumatic, and the vessel in either case may be
readily reached in actual practice, without attendance to
the somewhat complicated directions given for finding
it ; in the former case the swelling caused by the position of the sac, and in the latter the direction of the
already existing external wound, would guide the surgeon.

The internal pudic artery lies very deep in the gluteal
region, and having escaped between the pyriformis and
levator ani, emerges from the pelvis at the great sacro-
sciatic notch, and winds round the spine of the ischium
accompanied by its nerve ; again reaching the pelvis at
the lesser sacro-sciatic notch, it lies on the inner surface
of the ischium, and is there covered by a process of the
obturator fascia (*vide* Perineum).

The great sciatic nerve emerges from the greater notch
at its lower portion, and lies exactly between the tuberosity of the ischium and the great trochanter.

The intermuscular, cellulo-fatty membranes, which
are so extensive in the gluteal region, freely intercom-
municate with those within the pelvis, and readily explain the passage of pus either from or into its cavity.
Sciatic herniæ occasionally exist, a portion of intestine
passing down through the greater sacro-sciatic foramen,
and having the vessels posterior to its sac.

The bursa over the great trochanter is sometimes the
seat of abscess, which may be mistaken for diseases of
the hip-joint.

The obturator or *ischio-pubic* region forms the inner

boundary of the region of the hip, and comprises the obturator foramen, the structures covering it on either surface, and the parts immediately adjacent. The bony prominences are easily felt; they are the pubic space and horizontal ramus of the pubes anteriorly, and the descending ramus and tuberosity of the ischium internally and behind. The anterior surface consists of the gracilis, pectineus and adductors, beneath which is the obturator internus. The obturator foramen is not entirely filled in by the membrane, being wanting just below the groove through which the obturator artery and nerve pass. Attached to the inner or pelvic surface of the membrane and to the bone around, is the obturator internus, having the obturator fascia below, between it and the levator ani, and the peritoneum above; it is intimately connected with the margins of the obturator membrane, and assists in forming the *obturator canal*. This obturator canal is about three-quarters of an inch in length, and directed obliquely from above downwards, and from without inwards.

Obturator Hernia.—Occasionally a hernia protrudes through the canal, and forms a swelling in the adductor or pubic region, emerging upon the thigh, below the horizontal ramus of the pubes, to the inner side of the capsule of the hip-joint, having the femoral vessels in front, and a little to the outer side, the tendon of the adductor longus to the inner side, and behind, the pectineus muscle. The obturator vessels and nerve also pass through it to the thigh, and the pressure exerted by the hernia upon the nerve induces pain in those regions to which it is supplied, and is a fact of importance in diagnosis.

SURGICAL ANATOMY OF THE HIP-JOINT.

The hip-joint itself lies inclosed in the foregoing regions, which have been described from the surface inwards; and before entering upon a description of the joint itself, it is important to examine the relations of the muscles to the articulation, and their control over its movements, in order to understand the exact position of the head of the femur in the several dislocations to which the joint is liable, and their action upon the upper portion of the thigh bone in fractures.

Immediately in *front* of the joint is the tendon of the psoas and iliacus (separated from the capsular ligament by a bursa); *above* is the reflected tendon of the rectus femoris and the gluteus minimus, closely interwoven with the capsule; *internally* the obturator externus and pectineus; *posteriorly* the pyriformis, obturator internus, and gemelli, tendon of obturator externus, and quadratus femoris (*vide* Fig. 40). All these muscles are in absolute relation with the capsular ligament, and are covered in by the superficial muscles already described.

Being an enarthrodial joint, the movements of which it is capable are very extensive: *Flexion*, which is produced by the psoas and iliacus, sartorius and rectus femoris. *Extension*, by the hamstrings and some fibres of the gluteus maximus. In both these movements the neck of the femur rotates on its axis, whilst the inferior extremity of the thigh bone describes an arc of a circle, directly backwards and forwards. *Adduction* is performed by the pectineus, adductors and gracilis. In this movement the shaft of the femur is adducted to the middle line of the body, and its neck is lowered. *Abduction*, by the gluteus medius and minimus, and the tensor vaginæ femoris. The neck of the femur is raised. *Rotation outwards*, in which the trochanter major is thrown backwards and the foot outwards, by the gluteus maximus and medius, pyriformis, obturators, and quadratus femoris. *Internal rotation*, when the great trochanter is thrown forwards and the foot inwards, by the anterior fibres of the gluteus medius and the gluteus minimus.

It will be seen that the greater part of these muscles are external rotators, and it has been considered that this fact explains the corresponding rotation of the thigh in fracture of the neck of that bone, but it is much more probable that the eversion of the limb is due simply to its own weight. External rotation, however, is of no diagnostic value in fracture of the neck, unless accompanied by actual shortening of the limb.

The *trochanter major* is the great lever into which are inserted the rotator muscles of the hip-joint, and is separated from the integuments by the anterior edge of the tendon of the gluteus maximus, beneath which is a large bursa. Its variable position is readily seen in the several movements of the hip, describing the arc of a circle, owing to its continuity with the neck of the femur. When the neck of the femur is fractured, the trochanter major rotates in the axis of the bone, and in cases of dislocation, or suspected dislocation, measurements of its distance from the anterior superior iliac spine must be carefully compared with similar measurements taken on the opposite side of the body. The position of the *trochanter minor*, which lies just below the neck, at the superior and inferior aspect of the femur, is of importance to the surgeon in performing amputation at the articulation, as the knife is liable to be locked in it unless care be taken to pass it well behind.

242 SURGICAL ANATOMY OF

The capsular ligament is the thickest and strongest in the body, and particularly that anterior portion known variously as the ilio-femoral band, or the inverted Y-shaped ligament of Bigelow, of which the tail of the Y is attached to the anterior inferior spinous process of the ilium, and the fork to the root of the great trochanter and intertrochanteric line. It has been shown by Bigelow that this thickened portion of the capsule is the chief agent in producing the characters of the *regular* varieties of luxation. When it is ruptured in dislocation, it is almost always at its base, and so strong is it in some instances, that the margins of the cotyloid cavity have given way.

The neck of the femur varies as regards its obliquity to the shaft with the age of the individual. Before puberty it is very oblique, and almost in a continuous axis with the shaft; in the adult male it is at an obtuse angle with it, and directed upwards, inwards, and forwards, whilst in old persons it becomes horizontal.

The head of the femur presents a smooth ball, of very regular form, somewhat more than hemispherical, directed upward, inward, and a little forward, for articulation with the acetabulum. It has a separate point of ossification, becoming united to the shaft at the eighteenth year.

The *synovial membrane* covers all that portion of the neck within the joint, and is reflected on to the internal surface of the capsule, ensheathing the ligamentum teres, and often communicating anteriorly with the bursa beneath the psoas and iliacus.

The cotyloid cavity is deepened by the cotyloid ligament, rendered continuous below by the transverse ligament, beneath which the nutrient vessels pass to the joint.

The *ligamentum teres* is attached by its apex into a fossa just behind and below the centre of the globular head of the bone, and by its base to the margins of the notch at the bottom of the acetabulum, and its office is to check external rotation and adduction when the thigh is flexed, thus assisting in the prevention of dislocation of the head forwards and outwards. The bottom of the cotyloid cavity is very thin, and is liable to perforation in caries, in which case the pus invades the pelvic cavity. It may be opened by the point of the knife in amputation through the hip-joint, or in the subsequent gouging after resection, unless care be taken. It is a somewhat remarkable fact that after the operation of disarticulation, or in the case of old unreduced dislocation, the cavity contracts.

The *vessels which supply the articulation* are the obturator, ischiatic, internal circumflex, and gluteal; and the *nerves* are from the great sciatic, obturator, and accessory obturator, and they enter it either by means of the notch, or through the ligamentum teres.

Dislocation of the Head of the Femur.—With regard to the displacements of the head of the femur in dislocation, none of the muscles of the gluteal region probably exert any influence excepting the obturator internus, the muscular substance of which is so mixed with tendinous structure as to give it great strength; and when in a state of contraction it may be regarded as an accessory ligament to the joint.

The *regular dislocations* of the head of the thigh bone are—(1) upwards, on to the dorsum ilii; (2) backwards, into the great sciatic arch; (3) downwards, into the obturator foramen, and (4) forwards, on to the pubes. In the *first* form the limb is shortened from one to two and a half inches, and slightly bent, the knee resting on the opposite thigh, and the great toe upon the opposite instep, the head of the femur being felt beneath the glutei. In the *second* form the limb is shortened for about half an inch, and the thigh turned inwards and slightly flexed, the ball of the great toe lying on the base of the metacarpal bone of the opposite foot, and the head of the thigh bone protruding below and behind the tendon of the obturator internus. In the *third* form, the limb is lengthened for about one or two inches, the thigh is flexed, and abducted and advanced in front of the opposite one, the toes pointing downwards and forwards, and the trunk flexed on account of the tension on the psoas and iliacus muscle. In the *fourth* form the limb is shortened, rotated outwards, and the head of the bone felt on the pubes, just below Poupart's ligament. The limb, moreover, is abducted, and the foot points directly outwards.

The action of the muscles is well marked in cases of *fracture,* either of the neck of the femur internal to the capsule, or just below the trochanter, as far as the upper fragment is concerned. In the former case, which is *the* fracture of old age, and is a result of the slightest mishap, there is eversion of the limb, produced, according to some authorities, by the action of the external rotators, but far more likely by the weight of the foot alone, and shortening, which is produced by the action of the glutei, rectus, and hamstrings. In the case of fracture below the trochanters, a result of direct violence, the upper fragment is pulled forwards by the psoas and iliacus, everted and drawn outwards by the external rotators. There is shortening of the limb beyond the point of fracture, owing to the action of the rectus in front and the hamstrings behind, and the upper end is thrown outwards and the lower inwards, and everted by the adductors.

The most important operations in this region are *amputation through the hip-joint*, and *excision of the head of*

Fig. 41.

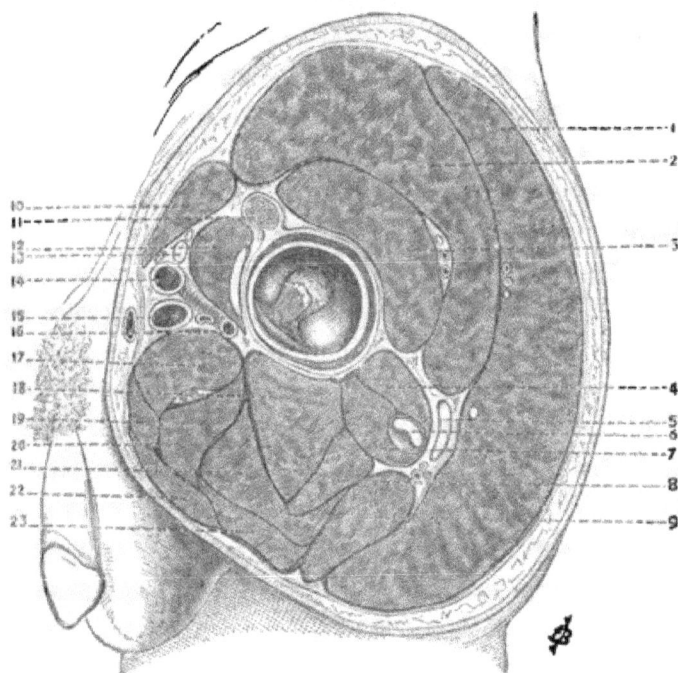

Section through the hip and gluteal region. 1. Gluteus maximus. 2. Gluteus medius. 3. Gluteus minimus. 4. Pyriformis. 5. Great sciatic nerve and ischiatic vessels. 6. Obturator internus. 7. Gemelli. 8. Biceps. 9. Quadratus femoris. 10. Sartorius. 11. Reflected tendon rectus. 12. Psoas and iliacus and bursa. 13. Anterior crural nerve. 14. Common femoral artery. 15. Common femoral vein. 16. Profunda vessels. 17. Gracilis. 18. Semi-membranosus. 19. Adductor brevis. 20. Semi-tendinosus. 21. Obturator externus. 22. Adductor longus. 23. Adductor magnus. (Altered from Béraud.)

the femur. In the operation of amputation, supposing the most rapid method, namely, that by anterior and posterior flaps, be performed, the anterior flap can be cut, and the articulation opened by the first thrust of the knife, if a point midway between the anterior superior spinous process of the ilium and the trochanter major be taken for entering the instrument, and its point be insinuated onwards towards the capsule, transfixing it, and directed downwards, forwards, and inwards, to a point just in front of the tuberosity of the ischium; the knife is then to cut the anterior flap from the front of the thigh, and an assistant is to help the operator by rotating the limb forcibly *outwards*, so as to bring the globular head of the femur and its attached ligamentum teres towards the surface; the ligament and capsule being divided, the assistant rotates the limb *inwards*, at the same time drawing it towards himself, so that the operator's knife, after cutting through the posterior portion of the capsule, may clear the trochanter major, and fashion the hinder flap.

Structures Divided in Amputation through the Hip-joint by the Antero-posterior Flaps.—In the *anterior*, integument, fasciæ, and superficial vessels, sartorius, anterior crural nerve, femoral vessels, rectus, tensor fasciæ, iliacus and psoas, portion of gracilis, adductor longus, adductor brevis, pectineus, and profunda vessels, and part of obturator externus and glutei, with superior gluteal and external circumflex vessels and nerves, capsular ligament, and ligamentum teres. In the *posterior*, part of the gracilis, adductor longus and brevis, and the adductor magnus and pectineus, internal circumflex artery, obturator nerve, quadratus femoris, part of obturator externus, obturator internus and gemelli, the hamstring muscles, sciatic vessels and nerves, part of gluteus minimus and medius, and the gluteus maximus, with its vessels, and the integuments of the buttock.

Excision of the Hip-joint.—The joint may be exposed either by a straight, curved, or T-incision, according to

circumstances; the straight portion should commence
just below the anterior superior spinous process of the
ilium, and be carried vertically over the trochanter
major, and the following structures divided: the gluteus
medius and minimus, obturator internus and gemelli,
obturator externus, pyriformis muscles, and the capsular
ligament. The head of the bone is protruded through
the wound by bringing the knee of the affected side
forcibly across the opposite thigh, with the toes everted.
The bone is to be divided below the level of the trochanter major. A more scientific proceeding is to separate the periosteum entire from the trochanter, leaving
the attachment of the muscles; by this means they retain
in a great measure their proper action, and, moreover,
new bone is thrown out.

SURGICAL ANATOMY OF THE MIDDLE FEMORAL REGION.

The limits of this region may be indicated *superiorly*
by a line drawn round the thigh at the fold of the nates,
and *inferiorly* by one drawn round the lower portion of
the thigh at about an inch above the patella; it has
the form of a truncated cone, with the base directed upwards.

Surface Markings.—*Anteriorly*, the prominences of the
oblique crossing of the sartorius, extending from the anterior iliac spine to the inner side of the knee, and of the
tensor fasciæ femoris, inclosing a triangular interval, in
which is the commencement of the quadriceps extensor,
which forms two curved muscular masses, fuller inferiorly, and inclosing a small triangular interval immediately above the patella, corresponding to its tendon of

insertion. *Posteriorly*, the surface is convex, and inferiorly is seen the divergence of the muscular masses which form the popliteal space. *Externally*, the surface is convex, and separated from the posterior by a deep furrow, marking the position of the external intermuscular aponeurosis. *Internally* and superiorly is the superior femoral region and Scarpa's space; the middle of this surface is flat, and indicates the position of the crossing of the sartorius; inferiorly is a large oval eminence, due to the vastus internus.

The *course of the femoral artery*, which is seen pulsating in Scarpa's space, is indicated by a line drawn from the centre of Poupart's ligament to a point just behind the internal condyle of the femur.

Dissection.—The **skin of** this region is very thick posteriorly, thinner **in front** and internally, and freely **supplied** with sebaceous glands in the upper and inner portion near the groin.

The subcutaneous cellular tissue generally contains a good deal of fat, and the superficial veins, nerves, and lymphatics lie in it. The most important superficial vessel is the *internal saphena vein;* its course in the thigh commences just behind the posterior part of the internal condyle, and passing obliquely upwards perforates the fascia **lata at the** saphenous opening, being there invested by the cribriform fascia, and terminates **in the** common femoral vein. It is very subject to a varicose condition, and may become enormously dilated.

The course of the *lymphatics* is pretty much that of the vein, and they terminate in those lymphatic ganglia, situate in Scarpa's space, which lie in the axis of the thigh.

The *superficial nerves* are derived from the anterior

crural, ilio-inguinal and crural branch of genito-crural
nerves; and posteriorly the integument is supplied by the
lesser sciatic.

Beneath the subcutaneous cellular tissue is the *fascia
lata*, forming an envelope for the muscles, the arrangement of which in the upper portion of the thigh has been
described with those regions. A very strong dense process, into which the tensor fasciæ is inserted, is attached
to the head of the fibula and to the outer surface of the
knee-joint. This fascia invests the muscles so closely
and firmly that its rupture allows of the bulging of the
fibres of the subjacent muscles to an extent which would
hardly be credited unless seen. Processes of this enveloping fascia form special sheaths for the muscles.

The fascia lata is attached deeply, on the outer side,
to the line leading from the trochanter major to the
linea aspera, and extends downwards to the tip of the
external condyle, and on the inner side to the line leading from the lesser trochanter to the linea aspera, and
downwards to the tip of the internal condyle. Thus it
will be seen, that the intermuscular septa thus formed,
divide the thigh into two distinct compartments, an anterior and a posterior.

An intercommunication, however, does take place,
owing to the passage of the femoral vessels through the
adductor opening, of the perforating branches superiorly,
and of the upper articulating branches of the popliteal
vessels inferiorly.

The *anterior* of these compartments, beneath the fascia
lata, contains externally and above the tensor fasciæ,
which is inserted obliquely into its substance; the sartorius, in its own sheath, which passes obliquely from
the anterior superior iliac spine, and wraps round the

thigh, being throughout its extent from the apex of Scarpa's space, a satellite to the femoral artery and vein; between these muscles lie the rectus femoris, becoming associated in the inferior third with the underlying muscles, the vasti and crureus, which envelop the femur from the great trochanter to the patella. Beneath the crureus is the subcrureus, inserted into the synovial membrane of the knee-joint, which extends upwards beneath the extensors and the periosteum of the femur for about three to four inches. Its office is to draw up the membrane, so that in extreme and sudden extension of the articulation it may not be pinched between the femur and the patella.

The Femoral Artery in the middle of the Thigh—Hunter's Canal.—Commencing at the apex of Scarpa's space, the artery describes an oblique course, lying covered over by the sartorius in its sheath, immediately beneath which is a strong fibrous aponeurosis, derived from the vastus internus externally, and the tendons of insertion of the adductor longus and magnus internally. This aponeurosis is very thin in the upper part of the middle femoral region, but becomes very dense lower down, terminating in a sharp margin, beneath which the internal saphena nerve leaves the vessel. External to the femoral vessels, is the vastus internus muscle; and internal to them are the tendons of the adductor longus and magnus, and behind them are the conjoined tendons of the vastus internus and adductors; and in the middle third, the fibres of the vastus internus alone separate the vessels from the femur. In this canal, which is triangular in section, with its apex at the femur, lie the femoral artery and vein posterior to and very intimately united with it; the long saphena nerve enters it with the vessels, above and

externally, and after crossing the artery, leaves the canal
at the point above indicated, and is distributed to the
skin of the knee and inner side of the leg.   The *anas-*

Fig. 42.

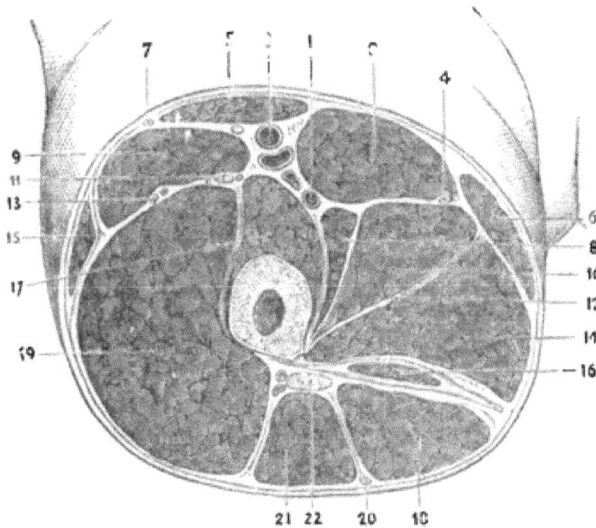

Section of the right thigh at the apex of Scarpa's triangle. 1. Profunda vessels. 2. Adductor longus. 3. Femoral vessels. 4. Superficial obturator nerve.
5. Sartorius. 6. Gracilis. 7. External cutaneous nerve. 8. Pectineus. 9. Rectus femoris. 10. Adductor brevis. 11. Anterior crural nerve. 12. Deep obturator nerve. 13. External circumflex vessels. 14. Adductor magnus. 15. Tensor
fasciæ femoris. 16. Semi-membranosus. 17. Vastus internus and crureus. 18.
Semi-tendinosus. 19. Vastus externus. 20. Small sciatic nerve. 21. Biceps
femoris. 22. Great sciatic nerve. (HEATH.)

tomotica magna artery is generally given off from the
trunk, just before the vessel becomes popliteal, that is,
before it passes through the adductor opening.

The superficial femoral is easily *compressed* against the
femur, at the middle of the inner third of the thigh.

Ligature of the Femoral Artery in Hunter's Canal.—
This operation is rarely performed nowadays, unless it
be for a wound in this portion of its course, the ligature

of the femoral for popliteal aneurism being applied in
Scarpa's space.

An incision is to be made in the course of the vessel
about three inches in length, through the integument
and fascia lata, until the oblique fibres of the sartorius
are recognized. Its edge reached, the muscle is to be
pulled upwards, when the aponeurotic fibres bridging
over the vessels will be seen. These are to be pinched
up and divided on a director, when the artery (and perhaps the internal saphena nerve) will be seen with its
vein, which is either behind it, or a little to its outer
side, and closely united to it by a dense fibrous investment. Occasionally the anastomotica magna is very
large and superficial, and may be mistaken for the main
trunk.

The *posterior compartment* of the thigh, as formed by
the fascia lata, contains—the hamstring muscles, the
great sciatic nerve, a great deal of fat and cellular tissue,
and the terminations of the perforating branches of the
deep femoral vessels; it presents but few points of surgical interest.

If amputation through the middle third were performed
by means of antero-posterior flaps (the operation to be
preferred), the **anterior** would contain the integuments
of the thigh, with the cutaneous nerves and internal
saphena vein, the rectus, sartorius, adductor longus,
brevis, and gracilis muscles, obturator nerve, femoral
vessels, and branches of anterior crural nerve, a portion
of the vastus externus, internus, and adductor magnus
muscles. The *posterior*, portion of vastus externus and
internus, adductor longus, brevis, and magnus muscles,
deep femoral and perforating vessels, the hamstring muscles, great sciatic nerve, lesser sciatic nerve, and integu-

ments of back of thigh.   If the amputation be performed near the knee, the muscles are liable to great contraction, since they take their fixed points at the pelvis superiorly. Hence the flaps, which should be lateral, external and internal, must be cut as long as possible.   The strong fibrous sheaths of the muscles favor the bagging of pus in amputations through the thigh.

In cases of *wounds* of the thigh, in which either the superficial or deep femoral is implicated, an approximative diagnosis of the site of the escape of blood may be made by examining the posterior tibial artery at the inner ankle.   If it pulsates, in all probability the superficial femoral is intact, and the profunda wounded; on the other hand, if the superficial femoral be the seat of injury, the blood, instead of continuing its course along it, and causing pulsation in the posterior tibial, is escaping into the surrounding tissues, and forming a false aneurism.   In enlarging the wound to find the bleeding point, it must be remembered that the profunda is external to the superficial femoral, and that the veins of either will be most likely involved.

In *fractures* of the shaft of the femur, the lower fragment is always drawn to the inner side of the upper one, and rotated outwards, although in some instances inwards: in the former case, owing to the contraction of the psoas and iliacus, and external rotators; in the latter, to the internal rotators.

SURGICAL ANATOMY OF THE REGION OF THE KNEE.

This region is limited below by a line drawn round the leg just below the internal tuberosity of the tibia ; it will be thus seen that the popliteal space, which partly belongs to the lower third of the thigh, the knee-joint, and the upper portion of the leg, is conveniently associated with the surgical anatomy of the knee.

Surface Markings.—If the leg be *extended* on the thigh, from above downwards, in front is the tendon of the quadriceps extensor, in which lies subcutaneously the patella, from the lower border of which descends the ligamentum patellæ passing to its insertion in the tubercle of the tibia; on either side of the quadriceps extensor tendon is a deep furrow, between it and the vasti muscles. In cases of synovitis, this furrow is obliterated, owing to the collection of fluid causing the synovial membrane to bulge beneath the tendon. If the leg be flexed, the condyles of the femur, and the interspace between them, are very evident, and the patella fills up the interval between the femur and the tibia, tuberosities of which are readily seen. In front of the patella and ligamentum patellæ the integument is slightly raised at a spot corresponding with the bursa patellæ.

Posteriorly, during complete extension the surface of the popliteal space is convex, and the positions of the muscles which bound it, although evident, are not so pronounced as when flexion commences; the most salient tendon is that of the semi-tendinosus. The position of the external and internal popliteal nerves lying in the middle of the space is easily seen during *extension*, as, being put on the stretch, they lie immediately beneath the integument, and resemble tendons. *Externally,* the knee presents a depression, formed by the obliquity of the axes of the femur and tibia, the deepest part of which corresponds with the position of the outer interarticular fibro-cartilage. Above this is the external condyle of the femur, below the external tuberosity of the tibia, whilst posterior to it is the head of the fibula and tendon of biceps; just behind the tendon of the biceps, and below the head of the fibula, can be felt the external popliteal nerve. *Internally* can be felt the internal condyle and internal tuberosity of the tibia, separated by the position of the interarticular fibro-cartilage. The internal saphena vein is seen beneath the integument, just behind the inner condyle.

The skin of the region is very thick and dense, and the subcutaneous cellular tissue contains, the internal saphena vein and nerve; on the inner side and in front of the patellæ, the bursa patellæ, effusion into which constitutes the affection known as "*housemaid's knee.*" In the early stage, inflammation of this bursa is to be distinguished from *synovitis* by the fact of the patella being hidden by the distended sac, whilst the cavities on either side of it remain; the fibrous tissue, however, along the edge of the patella being thin, pus may find its way into the synovial cavity. The aponeurosis is a continuation of the fascia lata, and is attached firmly to the tubercle of the tibia and its tuberosities, to the head of the fibula, afterwards blending with the fascia lata of the leg. Beneath this aponeurosis lie the muscles, which have the following relations :

Internally are the sartorius, gracilis, and semi-tendinosus, the tendons of which muscles strengthen the aponeurosis by their fibrous expansions, and are separated from the tibia by a large bursa, and the semi-membranosus. *Externally*, the tendon of the biceps, *anteriorly*, is the tendon of the quadriceps extensor and the ligamentum patellæ, between which and the tibia is a bursa (*bursa of Cloquet*). The articular branches of the popliteal, anterior tibial recurrent, and anastomotica magna ramify on the capsule.

The popliteal space forms the posterior aspect of the knee.

Dissection.—The skin is thin, and has beneath it a considerable amount of fat. It contains a number of lymphatic glands, which are liable to suppuration, or to enlargement after injuries to the foot or leg; they are divided into two series—a superficial, which accompany the saphena veins, and a deeper, which lie with the popliteal vessels; suppuration in these glands has been mistaken for aneurism.

The external saphena vein lies in this tissue before it perforates the popliteal aponeurosis to join the popliteal vein; here also are some branches of the small sciatic nerve. The popliteal aponeurosis continuous with the fascia lata above and fascia of leg below, is attached to the bony prominences and ligaments, and forms sheaths for the muscles and vessels of the space; strong transverse bands stretch across the space, and by connecting the tendons of the hamstrings, the fascia is rendered very tense. This fascia, from its strength and power of resistance, complicates the diagnosis of tumors in the space.

The boundaries of the popliteal space are, externally, the biceps above, the external head of gastrocnemius and origin of plantaris below. *Internally,* the tendons of the semi-tendinosus, semi-membranosus, gracilis and sartorius above, and the inner head of the gastrocnemius below.

These muscles are very subject to contraction after strumous disease of the knee-joint, and to cause subsequent dislocation of the leg upon the thigh, in cases where the disease has not been combated by treatment.

Connected with the tendons of these muscles are bursal sacs, which are of surgical importance, as if enlarged or inflamed they may offer some difficulties in diagnosis. Thus, one exists between the inner head of the gastrocnemius and the condyle of the femur, and often communicates with the joint, another exists between the tendon of the semi-membranosus and the tibia; the bursa beneath the outer head of the gastrocnemius is generally a prolongation of the synovial membrane of the joint, and between the popliteal tendon and the posterior and external lateral ligaments there often are found separate bursæ.

The muscles above mentioned inclose a lozenge-shaped space, containing a large quantity of fat and cellular tissue, in which lie the popliteal vessels, nerves, and some lymphatics; as this fat and cellular tissue is continuous with that surrounding the muscles of the back of the thigh and calf, any collections of pus in the space are liable to extend up the limb or downwards amongst the muscles of the back of the leg. Most superficial in the space, on the outer side, is the external popliteal nerve, which lies on the inner margin of the biceps tendon, and must be carefully avoided in tenotomy; the external saphena vein lies to its inner side, after having perforated the popliteal aponeurosis. More internal still is the internal politeal nerve, which is the continuation directly downwards of the sciatic nerve, in the inferior portion of the space; this nerve gives off a leash of branches which supply the muscles of the calf, and a filament, the communicans poplitei, which joins a corresponding one from the external popliteal, the communicans peronei, forming a loop which generally lies in the sulcus between the two heads of the gastrocnemius (*external saphena*). The sheath of the popliteal vessels is very dense, and incloses the *popliteal artery and vein*, which latter lies internal to the internal popliteal nerve, and superficial to the artery.

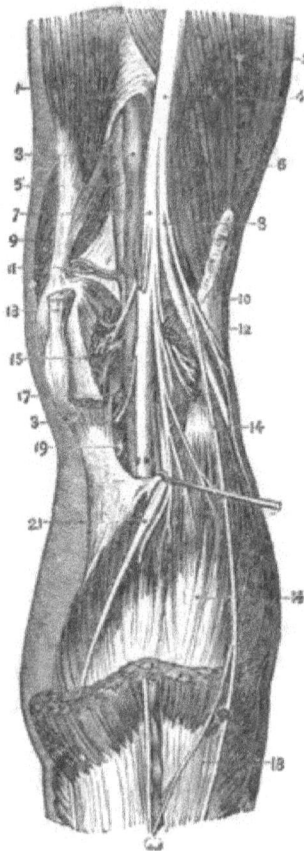

FIG. 43.

The Popliteal Artery.—The course of this vessel in the space is indicated by a line which, commencing at the centre of Poupart's ligament, and wrapping round the thigh, would fall immediately between the two condyles of the femur behind; commencing at the opening in the adductor magnus, it extends to the lower border of the popliteus muscle; it lies close to the surface of the bone, and gives off its articular branches nearly at right angles to its course. Between the artery and the vein is the articular branch of the obturator nerve, which supplies the knee-joint. The artery and vein are in such intimate relation that it would be almost impossible for a

Deep dissection of the popliteal space. 1. Adductor magnus. 2. Vastus externus. 3. Popliteal vein. 4. Great sciatic nerve. 5. Popliteal artery. 6. Short head of biceps. 7. Internal popliteal nerve. 8. External popliteal nerve. 9. Vastus internus. 10. Long head of biceps (cut). 11. Superior internal articular artery. 12. Outer head of gastrocnemius. 13. Tendon of semi-membranosus. 14. Communicans peroni nerve. 15. Inner head of gastrocnemius. 16. Soleus. 17. Inferior internal articular artery. 18. Gastrocnemius. 19. Popliteus. 20. External saphenous vein and nerve. 21. Tendon of plantaris. (HEATH.)

REGION OF THE KNEE.

punctured wound of the ham which entered the artery, not to involve the vein also.

Relations of the Popliteal Artery.—*In front*, just beneath the tendinous arch in the adductor magnus tendon, is the inner side of the femur; having wound round that bone, it has that portion of it between the bifurcation of the linea aspera in front of it, in the middle of its course the posterior ligament of the articulation, and below, the popliteal fascia. *Behind*, is the popliteal vein, internal popliteal nerve, aponeurosis; *externally*, biceps; *internally*, semi-membranosus.

The branches of the vessel are the muscular, the superior and inferior external and internal articular, and the azygos, which pierces the posterior ligament. These vessels maintain a very free anastomosis round the joint amongst themselves, the anastomotica magna, anterior tibial recurrent, and muscular branches.

Ligature of the popliteal, as a definite operation, is never practiced in modern surgery, for reasons which will be found discussed in works on aneurism.

The *diagnosis of tumors* in the popliteal space may, in most instances, be reduced to anatomical principles, thus: such tumors must be either aneurism (circumscribed or diffused), abscess, enlarged glands, cysts, or growths, &c. In the event of an aneurism, compression of the femoral would empty its sac, and the sound communicated to the ear by a stethoscope would be of a prolonged, blowing nature. Cysts, such as those alluded to as connected with the tendons, even if they have a communicated pulsation from being so closely packed together with the vessel, could be dragged away from it, and then these seeming pulsations would cease. In the diagnosis between an abscess and an aneurism the sac of which had suppurated, or between an abscess and ruptured popliteal artery, forming a diffuse aneurism, difficulty might be expected; but here the condition of the pulse below will generally determine the case.

SURGICAL ANATOMY OF THE KNEE-JOINT.

The articulations of the knee-joint are three in number—viz., between the femur and the patella, between the femur and the tibia, and between the tibia and the head of the fibula. The structure of the bones entering into the formation of the joint is the same in each instance,—cancellated tissue, inclosed in a layer of compact tissue. The articular extremities of the bones are well supplied with bloodvessels, these enter the patella on its anterior, and the femur on its posterior surface.

The articular surface of the patella is divided vertically by a crest into two facets, the rounder of which corresponds with the outer, and the longer and flatter with the inner articular portion of the trochlea.

The lower articular end of the femur is convex in front and concave posteriorly, and the internal condyle is lower than, and a little posterior to, the external, presenting on its inner side the inner tuberosity for the attachment of the internal lateral ligament, and a well-marked tubercle, very plainly felt beneath the integument, for the attachment of the tendon of the adductor magnus. The tuberosity of the outer condyle is less prominent than that of the inner, and gives attachment to the external lateral ligament. The articular surface extends higher on the outer side than on the inner, and is moreover in advance of it and broader. During complete flexion only, the patella occupies the centre of the trochlea, but in extension it overlaps the outer portion of the articular surface, and rises, if the extension be extreme, half its diameter above it.

The femur and the patella are united by the fibrous capsule and by the tendon of the quadriceps extensor. The great power exerted by the action of the quadriceps on the patella above, and its strong attachment to the tibia by the ligamentum patellæ below, explain how this bone may be fractured transversely by muscular action.

Dislocations of the Patella.—This bone is most frequently dislocated *outwards.* The bone lying above and external to the external articular surface of the trochlea, the outward traction of the extensor muscles favors this form of displacement. Dislocation inwards is very rare. Dislocation with the bone lying vertically either on the outer or the inner condyle is also very rare.[1]

The Articulation between the Femur and the Tibia.— The inner articular surface of the condyle of the femur is less prominent anteriorly than the external, but is set lower and extends a little further backwards, and is more oblique laterally. These surfaces are separated posteriorly by the intercondyloid notch. The head of the tibia presents two concave articular surfaces, the external of which is the rounder, separated by the spine, in front and behind which is a rough depression to which are attached the crucial ligaments.

The femur and tibia are united by the capsular ligament, posteriorly by the posterior ligament, which is a prolongation upwards of the tendon of the semi-membranosus, an internal and external lateral ligament, two

---

[1] *Vide* case in practice of author, British Medical Journal, December, 1872.

crucial, the anterior or external and the internal or posterior, and the two interarticular semilunar fibro-cartilages.

The synovial membrane of the knee-joint ascends upwards beneath the extensor muscles, as a pouch, for
about three inches, and is reflected from the articular
surfaces of the femur, to the crucial ligaments and ar-

Fig. 44.

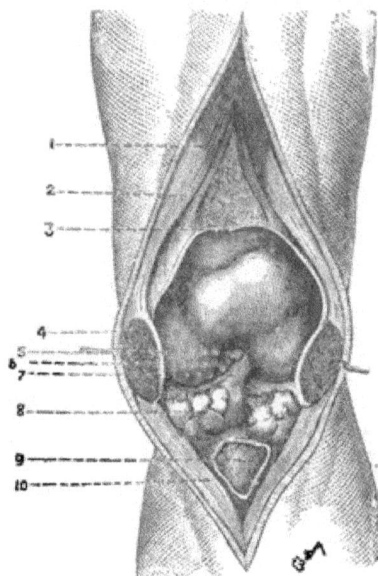

Knee-joint opened vertically. 1. Tendon of quadriceps extensor. 2. Subcrureus. 3. Cut edge of synovial membrane. 4. Patella divided vertically.
5. Ligamentum mucosum. 6. Posterior crucial ligament. 7. Anterior ditto.
8. Adipose tissue. 9. Bursa beneath ligamentum patellæ. 10. Ligamentum
patellæ.

ticular surface of the tibia, enveloping the semilunar
cartilages, and at the back of the external, forms a pouch
between its surface and that portion of the tendon of the
popliteus which is within the capsule; it then lines the
capsular ligament.

The lower epiphysis of the femur presents an ossified nodule at the ninth month of fœtal life, a fact of considerable value medico-legally in determining the age of the child. The entire epiphysis, however, does not unite with the shaft until the twentieth year; the upper epiphysis of the tibia at the twenty-fifth year.

Movements of the Knee-joint.—*Flexion* is performed by the biceps, semi-tendinosus, membranosus, popliteus, and accessorily by the gastrocnemius. *Extension* by the quadriceps extensor, and tensor vaginæ femoris. When the leg is semiflexed, the joint can be internally *rotated* by the sartorius, semi-tendinosus, and gracilis; *externally* by the biceps.

Relations of the Knee-joint, externally.—The tendon of the biceps and the strong process of fascia lata into which is inserted the tensor fasciæ; *internally* and a little *posteriorly*, the sartorius, semi-tendinosus, gracilis, and semi-membranosus; all these tendons are inclosed in bursal sheaths; *posteriorly*, the tendon of the popliteus and the tendon of the muscles forming the popliteal space, with the contents of the space itself.

Fracture of the patella when due to muscular action is always transverse, and the separation of the fragments to the upward traction of the quadriceps extensor acting on the upper one, whilst the lower is retained in position by the ligamentum patellæ. With a view to treatment, the tension should be relaxed, by extending the leg on the femur, and by slightly flexing the thigh on the pelvis so as to relax the rectus muscle. Non-union is sometimes owing to the bulging of the synovial membrane between the opposed fragments.

The articulation between the upper extremities of the tibia and fibula is an arthrodial joint, consisting of two

opposed articular spaces, united by two ligaments, an
anterior and posterior, with a synovial membrane between, occasionally communicating with that of the knee-
joint; a circumstance explaining the implication of the

FIG. 45.

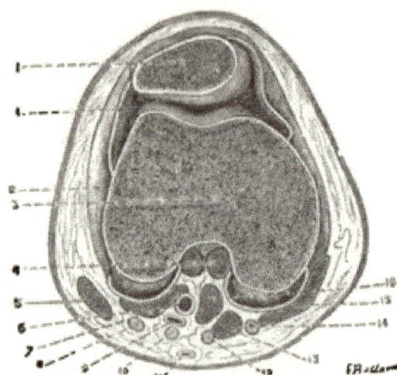

Horizontal section of knee-joint. 1. Patella. 1′. Synovial membrane. 2.
Capsule. 3. Femur. 4. Crucial ligaments. 5. Biceps. 6. Outer head of gastrocnemius. 7. Popliteal artery. 8. External popliteal nerve. 9. Popliteal
vein. 10. Internal popliteal nerve. 11. External saphena vein. 12. Semi-tendinosus. 13. Semi-membranosus. 14. Gracilis. 15. Sartorius. 16. Inner head
of gastrocnemius.

synovial membrane of the head of the fibula becoming
involved in effusions into the knee-joint.

The *operations* which concern the knee-joint are excision and amputation through it.

Excision of the Knee-joint.—The articular surfaces
which require removal may be exposed in several ways,
the simplest being by a semilunar incision, extending
from the inner side of the inner condyle to the outer side
of the external, the convexity of the incision lying midway between the lower border of the patella and the
tubercle of the tibia; the joint is thus opened at once.
The articular extremities of the femur and tibia, or pa-

tella, are to be removed according to circumstances; but in the instance of performing the operation on children, it is of great importance to avoid removing the entire epiphyses, as there would then be no further growth in the limb.

Structures Divided in Excision of the Knee-joint.—Integument and aponeurosis, patellar plexus of nerves, bursa patellæ, ligamentum patellæ, anterior part of capsular ligament, synovial membrane, crucial ligaments, lateral ligaments, articular vessels, and articular extremities of femur and tibia. The popliteal vessels are separated from the opened joint by the posterior ligament and popliteus muscle.

SURGICAL ANATOMY OF THE LEG.

The surgical region of the leg commences just below the knee, and extends to an imaginary line drawn round the lower part of the limb, just above the malleoli.

Surface Markings.—*Anteriorly* is the crest of the tibia, internal to which is the flat plane surface of the shaft of the bone, which being subcutaneous throughout permits of ready examination, and external to it is the mass of the tibialis anticus, and extensors of the toes. *Externally* are the peronei, separated by a well-marked groove corresponding to the interspace between them and the external edge of the soleus. *Posteriorly* is the swell of the calf, due to the gastrocnemius and soleus, the division between the two heads of the former being marked by a furrow continuous with the lower portion of the popliteal space. As the muscular fibres cease, the tendo-Achillis becomes more evident.

The leg can be readily divided into two regions, an

anterior and a posterior, limited by the inner border of the tibia internally and the outer border of the fibula externally.

Anterior Region.—The skin is freely movable over the subjacent tissues, and in the subcutaneous cellular tissue and fat lie the internal saphena vein, which crossing the inner malleolus, passes upwards toward the posterior border of the inner condyle of the femur, and is accompanied by the internal saphena nerve. The *aponeurosis* is the continuation downwards of the fascia lata, strengthened superiorly and internally by the expansion of the sartorius tendon, and thickened inferiorly where it forms the annular ligament of the ankle. It is adherent to the anterior surface of the tibia and external border of the fibula, and sends septa between and gives attachment to the anterior muscles of the limb; it is perforated in several places for the passage of the cutaneous nerves. Inclosed in the space between the aponeurosis superiorly and the tibia, fibula, and interosseous membrane posteriorly, lie in the first layer of muscles, the tibialis anticus and the extensor communis digitorum, united superiorly by an intermuscular septum lower down. They separate and inclose the origin of the extensor proprius pollicis, external to and below which is the peroneus tertius. The space between these muscles and the interosseous membrane is occupied by the anterior tibial vessels and nerve; the nerve pursuing the same course as the artery lies at first external to, then upon, and then again outside the vessels.

External Region.—The aponeurosis forms an investment for the peronei muscles; the compartment corresponding to the external surface of the fibula. To the upper and middle thirds of this surface the peroneus longus is attached; the upper fibres are pierced by the external popliteal nerve, which at this point divides into anterior tibial and musculo-cutaneous; the latter perforating the fascia about the middle third of the leg. Arising from the middle third of the fibula the peroneus brevis is inclosed in the same compartment.

The Anterior Tibial Artery.—The course of this vessel is indicated by a line drawn from the inner side of the head of the fibula to midway between the malleoli. It enters the region at a point below the popliteus muscle, and passes between the upper portion of the two heads of origin of the tibialis posticus, and comes off from the popliteal almost at a right angle. Its relations are, *anteriorly*, integument and fasciæ, tibialis anticus (above), extensor longus digitorum and extensor proprius pollicis, and the anterior tibial nerve; *internally*, tibialis anticus, extensor proprius pollicis (which crosses it at the instep); *externally*, the anterior tibial nerve, extensor longus digitorum, and extensor proprius pollicis; *posteriorly*, the interosseous membrane, the tibia, and anterior ligament of ankle-joint.

Ligature of this vessel is rarely required, unless it be for a wound, which would be enlarged, and the bleeding point sought for. To tie it, an incision should be made in the upper third, in the interspace between the tibialis anticus and the extensor communis digitorum; the intermuscular septum between them looked for, the muscular fibres detached from it, and pulled on one side, when the vessel, surrounded by venæ comites, and having its nerve to the outer side, will be seen lying on the interosseous membrane.

In the *lower* third, an incision should be made along the outer border of the tibialis anticus tendon, when the vessel will be found between it and the tendon of the extensor proprius pollicis, and the nerve generally lying on it.

Posterior Region.—Beneath the integument and superficial fascia are the external saphena vein and nerve, and some branches of the musculo-cutaneous and internal

FIG. 46.

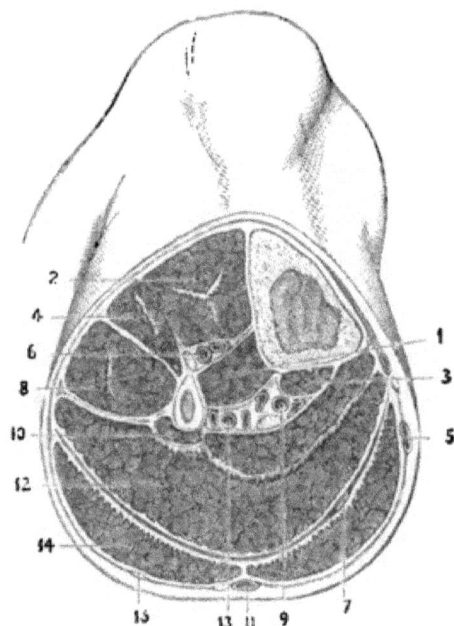

A section of the right leg in the upper third. 1. Tibialis posticus. 2. Tibialis anticus. 3. Flexor longus digitorum. 4. Extensor longus digitorum. 5. Internal saphenous vein. 6. Anterior tibial vessels and nerve. 7. Tendon of plantaris. 8. Peroneus longus. 9. Posterior tibial vessels and nerve. 10. Flexor longus pollicis. 11. External saphenous vein and nerve. 12. Soleus with fibrous intersection. 13. Peroneal vessels. 14. Gastrocnemius. 15. Communicans peronei nerve. (HEATH.)

saphena nerves. The aponeurotic sheath, inclosing that portion of the leg posteriorly between its attachments to the tibia and fibula, is subdivided by an expansion separating the superficial from the deep flexors, vessels, and

nerves. The most posterior contains the gastrocnemius
and soleus, uniting to form the tendo-Achillis, and the
plantaris, with a good deal of fat and bursal tissue.
The second, anterior to the former, contains the flexor
longus digitorum internally, the flexor longus pollicis
externally, and tibialis posticus muscles between them,
closely united by intermuscular septa. The posterior
tibial vessels on the tibial side, having the posterior
tibial nerve external to them, and the peroneal vessels
on the fibula, lying at first beneath the intermuscular
aponeuroses, next between the flexor longus pollicis and
tibialis posticus, and lower down the limb, between the
tibialis posticus and the fibula.

The posterior tibial artery would rarely require ligature in its upper third, unless for injury, in which case
the wound should be enlarged, and the bleeding point
secured; but if, however, the vessel be divided by a
punctured wound from the front of the leg, or in the
case of traumatic aneurism of the vessel, low down, it is
necessary to place a ligature upon it in this situation.

*Ligature of the Posterior Tibial Artery in the Upper
Third.*—This vessel is reached most scientifically by an
incision made along the posterior border of the subcutaneous surface of the tibia, about four inches in length,
dividing the integument and aponeurosis, taking care to
avoid the internal saphena vein and nerve. The inner
border of the gastrocnemius is to be drawn aside; when
the tibial head of the soleus is reached, its fibres are to
be divided, until the intermuscular septum (the position
of which is variable) is come upon. This is next to be
cut through, and the fibres of this muscle divided until
freedom of access is obtained. The cut edges of the
soleus are to be separated, the smooth intermuscular apo-

neurosis which separates the superficial from the deep flexors is to be divided on a director, and the posterior tibial nerve drawn on one side; the posterior tibial artery, surrounded by venæ comites, is seen lying on the flexor longus digitorum.

The nutritious artery is a branch of considerable importance. Directed upwards towards the knee, it enters the shaft of the tibia in a deep canal in the posterior aspect, about four fingers' breadth from the knee, and may give rise to troublesome hemorrhage in amputation of the leg at this part.

The peroneal artery, generally regarded as a branch of the posterior tibial, is very often of larger size. It is very deep, and lying along the fibular surface of the leg, has the following relations : *In front,* the tibialis posticus and flexor longus pollicis; *externally,* the fibula; and behind, the soleus, deep aponeurosis, and flexor longus pollicis. This vessel is occasionally wounded in compound comminuted fractures of the fibula.

In fractures of the tibia and fibula, which take place *obliquely* from above, downwards, and forwards, the muscles of the calf cause the lower fragments to be drawn upwards and backwards, and frequently the upper one to protrude through the integument. With a view of bringing the surfaces into apposition, the knee should be bent to relax the opposing muscles, and extension made from the knee and ankle (*vide* Ankle-joint).

Fracture of the lower end of the fibula is usually associated with fracture of the inner malleolus (Pott's fracture), or rupture of the internal lateral ligament of the ankle-joint, causing dislocation of the foot outwards. The eversion of the foot is due to the action of the peroneus longus, whilst the heel is drawn upwards by the gastrocnemius and soleus. The reduction is effected by flexing the leg at right angles with the thigh, and making extension from the knee and ankle.

The relation of the tibia and fibula to each other must be borne in mind in performing amputation through the leg. The fibula lies on a plane posterior to the tibia, and its external border, with about half of its external surface, is situated behind the interosseous ligament; hence, unless care be taken, the knife may be easily entered between the bones, instead of taking the necessary oblique course skirting their posterior surfaces.

Structures Divided in the Double Flap Amputation through the Calf.—In the *anterior* flap: the integument, cutaneous nerves, aponeurosis, tibialis anticus, extensor communis digitorum, and extensor proprius pollicis, peroneus longus and brevis, musculo-cutaneous nerve, anterior tibial vessels and nerve. In the *posterior*, the flexor longus digitorum, flexor longus pollicis and tibialis posticus, posterior tibial vessels and nerve, peroneal vessels, intermuscular aponeurosis, soleus and plantaris, gastrocnemius, external saphena nerve and vein, internal saphena vein and nerve, aponeurosis and integuments.

SURGICAL ANATOMY OF THE FOOT.

Ankle or Malleolar Region.—It has been thought more convenient to postpone the description of the several articulations entering into the conformation of the foot and ankle until all those soft structures which inclose them have been explained; as all surgical reference to the skeleton must necessarily be made from the surface, it is of importance that all the intermediate parts be de-

monstrated from without inwards, and in the order they
would be met with in an operation.

Surface Markings.—This region includes the ankle-
joint, and the structures immediately surrounding it,
and offers for examination two surfaces, an anterior and
a posterior.

Anterior Surface.—The two malleoli, of which the internal is the shorter and broader, and the external set
more backwards and longer, inclose a space through which
pass the extensors of the foot and toes, which are rendered evident in their several movements.   Beneath the
integument and superficial fascia lie internally, just in
front of the malleolus, the internal saphena vein, accompanied by its nerve ; more externally the musculo-cutaneous nerve, whilst passing from behind the outer malleolus is the external saphena nerve.   The aponeurosis
is a strengthened continuation of that of the leg, attached
intimately to the malleoli, consisting of a superior fasciculus, which binds down the subjacent tendons, just in
front of the extremities of the tibia and fibula, and an
inferior, which retains them in connection with the tarsus.
It forms two distinct septa, commencing internally—1,
for the tendon of the tibialis anticus ; 2, for the tendon
of the extensor longus digitorum, peroneus tertius, and
extensor proprius pollicis, beneath which is the anterior
tibial vessels and nerve.   These sheaths are lined with
separate synovial membranes.

The posterior surface, or that portion behind the
malleoli, is separated by the tendo-Achillis into two
hollows.   In the *outer,* beneath the integuments, is the
external saphena vein and nerve, lying upon the external
annular ligament, which is attached to the outer malleolus and outer surface of the os calcis, binding down the

peroneus longus and brevis, the former being the superior; they are contained in a common sheath (at first),
and have a common synovial membrane.

In the *inner*, the pulsations of the posterior tibial
artery are plainly seen or felt, and beneath the integument are the internal saphena vein and nerve, which lie
upon the internal annular ligament, which is attached
to the inner malleolus and inner surface of the os calcis,
and forms with the tibia, os calcis and astragalus a series
of separate canals, containing from before backwards
the tendons of the tibialis posticus; the flexor longus
digitorum; the posterior tibial vessels and nerve, running in a sheath of their own derived from the contigu-

Fig. 47.

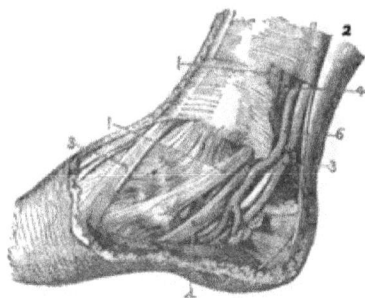

Relations of parts behind the inner malleolus. 1, 1. Tibialis posticus. 2. Tendo-
Achillis. 3. Tibialis anticus. 4, 4. Flexor longus digitorum. 6. Posterior tibial
artery. 8. Posterior tibial nerve. The tendon of the flexor longus pollicis is too
deeply placed to be shown in this view. (HEATH.)

ous septa; the flexor longus pollicis, the canal of which
is formed partly by the astragalus. Each of these canals
has a separate synovial membrane. The tendo-Achillis
has a separate sheath derived from this aponeurosis.

The posterior tibial artery at the ankle-joint lies between the flexor longus digitorum and the flexor longus pollicis tendons, having venæ comites on each side, and the posterior tibial nerve behind it.

Ligature of the Posterior Tibial Artery at the Inner Malleolus.—This vessel is easily reached, but the incision must be made carefully, as there is a risk of dividing it in overcoming the resistance of the internal lateral ligament. An incision about two inches and a half in length is to be made through the integument, midway between the inner malleolus and the tuberosity of the os calcis. After the dense aponeurosis is exposed it should be cautiously divided, when the vessel will be seen surrounded by venæ comites, and in order to avoid the nerve, which lies *posteriorly*, the needle should be passed *from the heel towards the ankle*.

The Dorsum of the Foot.—The chief points to be observed in the surface markings of the dorsum of the foot are those connected with the prominent points of its skeleton. For the performance of the several amputations and disarticulations, certain landmarks are necessary to guide the operator in finding the articulation he desires to open. Thus, a line drawn from the depression on the inner side of the foot, immediately between the inner cuneiform bone and the great toe, to the posterior edge of the tuberosity of the fifth metatarsal bone, indicates the course of an incision, such as would expose the tarso-metatarsal articulation.

Again, the tubercle of the scaphoid on the inner side, and a point midway between the outer malleolus, and the tuberosity of the fifth metatarsal bone, which is the situation of the articulation between the cuboid and os calcis, indicates a line of incision which would open the medio-tarsal joint.

The *structures met with in dissecting down* **upon the**
dorsal aspect of the tarsus and metatarsus, are—the integument, and subcutaneous cellular tissue, which contains the dorsal venous arch, the terminal inosculation
of the internal and external saphena, and the musculo-
cutaneous **nerves,** beneath which is the dorsal aponeurosis of the foot, and from within outwards the tendons of
the tibialis posticus, tibialis anticus, extensor proprius
pollicis, extensor communis digitorum, peroneus tertius,
and peroneus brevis, and in a plane beneath them the
extensor brevis **digitorum;** externally, **the dorsalis pedis**
vessels and anterior tibial nerve, the tarsal **and** metatar-
sal branches of the **anterior tibial artery and external**
branch of the anterior tibial nerve, which latter lie beneath the lesser extensor muscle. **All these structures**
lie close upon the tarsus and metatarsus, and **between
the** metatarsal bones the dorsal interossei **are seen bulging through.**

The *dorsalis pedis artery* is the continuation **of the**
anterior tibial, and passes forwards on the tibial side of
the foot to the inner interosseous space, where it divides
into the dorsalis hallucis and the perforating vessels
which enter the sole between the heads of the first dorsal
interosseous muscle, and is in *relation* in front with the
integument and fascia, and inner tendon of the extensor
brevis digitorum; internally with the extensor proprius
pollicis; externally with the extensor longus digitorum
and anterior tibial nerve; posteriorly with the astragalus, scaphoid, inner cuneiform, and with the ligaments
attached to them.

Ligature of Dorsalis Pedis Artery.—The course **of this**
vessel is indicated by a line drawn from the **middle of
the** intermalleolar space to the first interosseous space.

It is superficial, but is bound down by a very dense aponeurosis, which must be divided cautiously to avoid injuring the artery beneath.   An incision is to be made over the instep along the outer border of the extensor proprius pollicis, when the vessel will be found lying in a triangular interspace formed by the outer border of the extensor proprius pollicis internally, by the inner tendon of the extensor brevis externally, and by the fleshy fibres of the extensor brevis digitorum, posteriorly.   The nerve lies to its outer side.

SURGICAL ANATOMY OF THE SOLE OF THE FOOT.

Dissection.—On removing the integument, which is very thick and strong, the first tissue met with is a dense layer of fat, in which are three bursæ, one beneath the os calcis, and two beneath the heads of the first and fifth metatarsal bones.   Ramifying in the fat are some cutaneous branches of the cutaneous nerves of the foot, some perforating branches of the plantar vessels, and a great number of lymphatics.   The next layer is formed by the plantar fascia, consisting of three portions, of which the central is the strongest, sending down processes which inclose the several muscles, separating the middle from the external and plantar groups.   The fascia divides opposite the middle of the metatarsus into five processes, each one of which divides again opposite the metatarso-phalangeal joint into two slips, which by their deep attachments form arches for the passage of the flexor tendons to pass to the toes; the interspace allows of the digital vessels and nerves, the tendons of the lumbricales and interossei becoming superficial.   The mutual relations of the structures forming the sole of the foot can be conveniently referred to the partitions formed by the plantar fascia. *In the inner compartment*, internally and posteriorly, lie the fleshy fibres of the abductor pollicis, the tendons of the flexor longus digitorum and flexor longus pollicis, the latter crossing and becoming internal anteriorly, the posterior tibial vessels and nerves becoming plantar, flexor brevis pollicis, the vessels and nerves of the great toe, and inner side of the foot.

In the outer compartment lie the abductor minimi digiti, and the flexor brevis minimi digiti.

The middle compartment, which is by far the most important from its contents and size, is bounded by the plantar fascia below, laterally by the septa between the outer and inner compartments, and its roof is formed by the under surfaces of the bones forming the arch of the foot; it communicates posteriorly with the region of the leg by means of the sheaths for the tendons and vessels passing beneath the inner malleolus. Beneath the middle fasciculus of the plantar fascia, lie the flexor brevis digitorum, the plantar vessels and nerves, the flexor accessorius, the tendons of the flexor longus digitorum, with which are associated the lumbricales, and flexor longus pollicis internally. Beneath these muscles lie posteriorly the tarsal bones and their ligaments, anteriorly, the adductor pollicis, the heads of the metacarpal bones and the ligaments uniting them, the transversus pedis, the plantar arch and external plantar nerve and their interosseal branches, the bodies of the metacarpal bones, between which lie the plantar interossei, and lying deeply in the tarsus the tendon of the peroneus longus.

The Ankle-joint and Articulations of the Foot.—The structures immediately in relation with the ankle-joint are, *anteriorly,* the tendons of the extensor longus digitorum and peroneus tertius, the extensor proprius pollicis, anterior tibial vessels and nerve, the tendon of the tibialis anticus; *posteriorly,* tendons of the peroneus longus and brevis, the flexor longus pollicis, the posterior tibial vessels and nerve, the tendons of the flexor longus digitorum and tibialis posticus.

The joint is formed by the articulation of the inferior articular extremities of the tibia and fibula with the astragalus.

The Tibio-tarsal Articulation.—The extremities of the tibia and fibula are hollowed out into a sort of mortice,

Fig. 48.

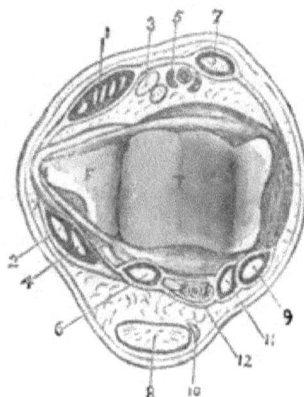

Section of the right ankle. 1. Extensor longus digitorum and peroneus tertius. 2. Peroneus longus. 3. Extensor proprius pollicis. 4. Peroneus brevis. 5. Anterior tibial vessels and nerve. 6. Flexor longus pollicis. 7. Tibialis anticus. 8. Tendo-Achillis. 9. Tibialis posticus. 10. Plantaris. 11. Flexor longus digitorum. 12. Posterior tibial vessels and nerve. (HEATH.)

concave from before backwards, open in front and behind, and shut in laterally by the malleoli. The use of the malleoli is to prevent lateral movements, and to restrict the motions of the joint to flexion and extension. The articular surface of the astragalus presents superi-

orly a surface convex from before backwards for the corresponding articular surface of the tibia; laterally are
two articular surfaces corresponding with those of the
malleoli, the external one being the larger. The *ligaments* are an external lateral, consisting of three fasciculi; an internal, or deltoid, radiating from the inner
malleolus, to be attached to the scaphoid, os calcis, and
astragalus; anteriorly are a few fibres closing in the articulation; posteriorly the joint is shut in by the transverse ligament of the inferior tibio-fibular articulation.
The synovial membrane invests the inner surface of the
ligaments and articular cartilages, and owing to the
great laxity of the anterior and posterior ligaments it
readily bulges through them, consequently the joint is
easily reached from either aspect, and in cases of synovitis the membrane usually protrudes in front between
the malleoli and again behind the external malleolus.

Dislocation at the Tibio-tarsal Articulation—that is to
say, dislocation of the entire foot from the bones of the
leg—is almost invariably associated with fracture of one
or other malleoli, which can be readily understood from
the shape and extent of motion allowed at the joint. It
is a result of the foot being twisted in running or walk-
ing, and may either be *outwards,* when the lower end of
the fibula is broken, the inner malleolus or internal lateral ligament torn; *internal,* when there is no fracture
of the fibula, but the lower end of the tibia is broken;
backwards, when both malleoli are broken, and the heel
projecting; and *forwards,* when the astragalus is thrown
in front of the tibia.

Excision of the Ankle-joint (tibio-tarsal).—The seat of
the disease can be reached by a variety of incisions; but
the object to be attained is to save the tendons, in order

that they may still fulfil their functions as far as possible, and an accurate knowledge of the parts in contact with the articulation is necessary. The integument only being divided by an incision which commences just above and behind the outer malleolus and extending across the joint to a corresponding point above the inner, the flap being dissected back, the peronei tendons are to be dislodged and the external lateral ligament divided. In order to obtain access to the joint on the outside, the lower end of the fibula is to be snipped off, and its connection with the tibia severed. Next, to get at the inner aspect of the articulation, the flexor communis digitorum and tibialis posticus tendons are to be dissected from behind the inner malleolus, and care taken to avoid the posterior tibial vessels and nerve. The lower end of the tibia can now by a wrench be dislocated through the wound; the diseased surfaces are then to be removed (Hancock). The diseased surfaces can be reached and removed by two lateral incisions (Barwell).

The articulation of the astragalus with the os calcis is one of great strength, owing to the interosseous ligaments which lie in the grooves of these bones; it is ruptured in cases of dislocation of the astragalus from the os calcis. This is by far the most important of the luxations of the tarsal bones, and may occur either forwards and inwards, or forwards and outwards, or backwards.

Amputation at the Ankle-joint.—The landmarks for the guidance of the knife are, in the first place, for the *anterior flap*, the two malleoli, which are to be united by a semilunar incision; and for the *posterior*, one cutting the sole transversely and a little obliquely forwards, and extending between the limits of the preceding incision;

the articulation is **opened, and the** lateral ligaments **di-**vided, the posterior part of **the** capsule cut through, **and** the os calcis sawn through **obliquely from** behind forwards and downwards. **The anterior** flap **is** dissected off the malleoli, which **are next** sawn off, and the cut surfaces approximated, **and the tendo-Achillis,** perhaps, **divided.** (Pirogoff.) Syme's amputation **consists** in removing **the os** calcis, and sawing off the ends **of the** malleoli. In **both** these operations care **must be taken not to wound the trunk** of the posterior tibial artery, **and to keep the** external and internal plantar vessels **as long as** possible.

Club-foot (*Talipes*).—The various deformities **of the** foot, occurring at **the** tibio-tarsal or medio-tarsal articulations, which are within the operative interference **known** as tenotomy, are—*talipes equinus,* **in which the heel is** raised so that the individual walks **on the ball of the foot, in** which case the tendo-Achillis **requires division ;** *talipes varus,* **or more precisely talipes equino-varus, where the heel is raised, the foot turned** inwards **for about the anterior** two-thirds—this **inversion** taking place at **the astragalo-scaphoid** and calcaneo-cuboid **articulations, its dorsal** aspect outwards, and **the inner** edge drawn up; the **tendons** requiring division **being,—**the tendo-Achillis, **and the tendons of the tibialis** posticus, anticus, and **flexor longus digitorum ;** *talipes valgus,* where the **inner ankle is towards the ground and the** outer edge of **the foot** turned **up; the tendons to be divided are,—the** peronei, **and the extensor longus digitorum, and the plantar** fascia ; *talipes calcaneus,* **where** the patient walks on the heel,—a case requiring **division** of the tendons of the tibialis anticus, extensor communis digitorum, extensor proprius pollicis, **and** peroneus ter-

tius.   There are several intermediate forms; this is owing
to the fact that the posterior tibial nerve supplies the
gastrocnemius and soleus, which terminate in the tendo-
Achillis, and the tibialis posticus.   Again, talipes equinus
and varus are usually associated, because the extensor
and peronei muscles are supplied by the anterior tibial
and musculo-cutaneous nerve.   Talipes calcaneus and
valgus are generally associated.

The tibialis posticus tendon is divided in *tenotomy*,
either above or below the ankle.   The point selected
above the ankle is on the posterior margin of the tibia
about an inch or so above the malleolus, where it lies in
the groove in its own sheath and in contact with the
bone; eversion of the foot tenses the tendon *below* the
ankle, at its insertion into the scaphoid.   The point is
just above the astragalo-scaphoid articulation, which is
immediately behind the first tuberosity met with in passing the finger along the inner side of the foot, starting
at the malleolus.   The tibialis anticus not being so confined as the preceding can be more readily put on the
stretch, as it passes over the lower end of the tibia in the
innermost compartment of the annular ligament.   It may
be also divided at its insertion into the inner cuneiform
bone, the position of which attachment may be ascer-
tained by passing the finger along the inner surface of
the foot, when it is just in front of the articulation of the
scaphoid with the inner cuneiform.

In performing the *tarso-metatarsal disarticulation*,
known as Hey's or Lisfranc's, the line of the joint may
be exposed, in the first place, by starting from the outer
surface of the foot from a point immediately behind the
tuberosity of the fifth metatarsal bone to a point which
may be indicated in one of the following ways: (1) If

a transverse line be drawn across the foot from the
tuberosity of the fifth metatarsal bone, it falls on the
inside of the foot, two-thirds of an inch *behind* the required spot; (2) in following the inner edge of the foot
from behind forwards, an inch in front of the malleolus
is the projection of the scaphoid; the joint is one inch *in
front* of this. The articulation of the first metatarsal
bone with the inner cuneiform is oblique from within
outwards, and about a quarter of an inch in front of the
third. The line of the joint is rendered irregular by the
jutting into the tarsus of the second metacarpal bone,
which is wedged in between the inner and outer cuneiform bones, its line of articulation lying about half an
inch behind the anterior articular surface of the inner

FIG. 49.

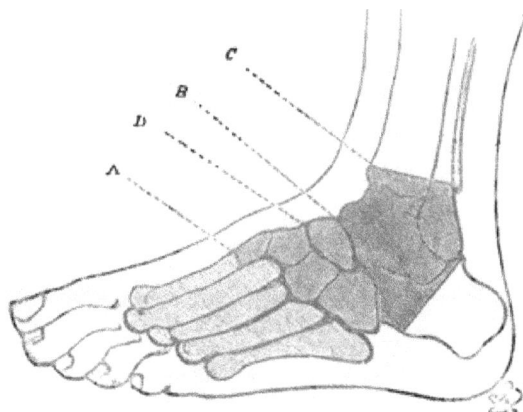

Rough sketch of portion of tarsus removed in—A. Hey's amputation.
B. Chopart. D. Roux. C. Pirogoff.

cuneiform, and about a quarter of an inch behind the
anterior articular surface of the outer cuneiform bone.
The joint of the third with the scaphoid is almost transverse; that of the fourth is curved; and that of the fifth

with the cuboid is doubly oblique. **After** disarticulation, the posterior **flap should** extend **as far as the** web of the toes.

Parts Divided in Hey's **Amputation.—In the** *anterior* flap, **the** integument beginning from **the outer** side, the dorsal veins of the foot, the internal and external divisions of the musculo-cutaneous nerve, the internal **and** external cutaneous nerves, the dorsal aponeurosis, **extensor** brevis digitorum, tendon of peroneus brevis, tendon of extensor communis digitorum, **anterior** tibial vessels **and nerve, tendons** of extensor proprius pollicis and **tibialis anticus,** dorsal ligaments, and the articulation.

In **the** *posterior* flap, plantar ligaments, tendon **of** peroneus longus, external and internal plantar vessels and nerves, interossei, the **flexor** brevis, abductor and adductor **pollicis,** transversus pedis, **tendons of long and** short flexors **of** toes, and flexor longus **pollicis** tendon, digital vessels and **nerves,** plantar fascia and integument.

Chopart's **amputation, or the** medio-tarsal, **consists of** opening the articulation **by a semilunar incision,** extending between the joint **behind the** tubercle **of** the scaphoid **internally** and a **point midway** between the external malleolus and the tuberosity **of the** fifth metatarsal bone; externally, the **posterior flap is to be brought well** up to the web of **the toes.**

It is important to **remember** that the direction of the articulating surface **is changed in flexion or** extension : in flexion the astragalus **and calcis** are in the same line, in extension the calcis is at least a **quarter of an** inch in front ; **the** head of the astragalus presents **a** large globular surface, whilst the anterior articulating surface of the calcis is concave.

The following directions for **discovering** the articula

tion with readiness are useful. *To find its internal side,*
follow the inner edge of the foot with the finger; the
first tuberosity met with is the scaphoid; the joint is immediately behind it. For the *external side,* pass the finger along the outer edge of the foot from the external
malleolus; the articulation is immediately in front of the
first tuberosity met with, which belongs to the os calcis.

Fig. 50.

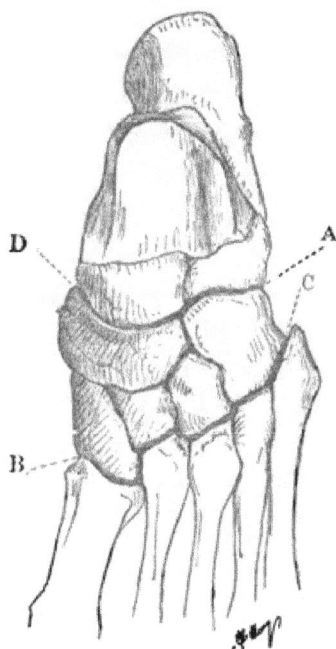

A. Line of Chopart.   B. Line of Hey.   c. Tubercle of fifth metatarsal.
D. Tubercle of scaphoid.

For its middle and superior portion, extend the foot and
abduct it; then, applying the finger on the union of the
external with the middle third of the intermalleolar
space, the first eminence met with in proceeding forwards

is the head of the astragalus; immediately in front of this is the articulation. (Malgaigne.)

Structures Divided in Chopart's Amputation.—Commencing from the inner side, the *anterior flap*, that is, up to the point of disarticulation, will contain integument, saphena vein, and musculo-cutaneous nerve, anterior annular ligament, tendon of tibialis anticus, extensor proprius pollicis, anterior tibial nerve and vessels, tendons of common extensor and peroneus tertius, extensor brevis digitorum, peroneus brevis and anterior ligaments of the articulation; the *posterior* flap should contain the posterior ligaments of the articulation, the tendon of the tibialis posticus, flexor longus digitorum and flexor longus pollicis, some branches of the internal plantar nerve and vessels, the abductor pollicis, the flexor accessorius, the tendon of the peroneus longus, the abductor minimum digiti, the flexor brevis digitorum, the external plantar nerve and vessels, and the integument of the sole.

Synovial Membranes of the Tarsus and Metatarsus.— There are four synovial membranes in the articulations of the tarsus—namely, one for the posterior calcaneo-astragaloid articulation, a second for the anterior calcaneo-astragaloid and astragalo-scaphoid, a third for the calcaneocuboid, and a fourth for the surfaces of the cuneiform with the scaphoid, the cuneiform with each other, the external cuneiform with the cuboid, and the middle and outer cuneiform with the second and third metatarsal bones. Between the internal cuneiform and the base of the metatarsal bone of the great toe there is a single synovial membrane, and there is another common to the anterior surface of the cuboid, and the bases of the fourth and fifth metatarsal bones.

The arteries of the sole of the foot are the internal and

external plantar; the internal commences at about the centre of the inner aspect of the os calcis, and courses forwards between the muscles of the inner and middle groups, anastomosing with the malleolar and dorsalis pedis. The external is much larger, and forms the plantar arch. Commencing at the same spot as the internal, it passes obliquely forwards and outwards, lying at first between the os calcis and abductor pollicis, and then between the flexor accessorius and flexor brevis, and forming a curve, the convexity of which is forwards, it joins, at the interval between the first and second metatarsal bones, the perforating branch of the dorsalis pedis, thus completing the plantar arch. Its branches are muscular, perforating, which inosculate with those of the metatarsal artery, and digital, which supply the three outer toes and half the second.

The nerves are the internal and external plantar branches of the posterior tibial, of which the internal is considerably the larger.

Toes.—The description already given of the fingers will in almost every particular suffice for that of the toes.

INDEX.

www.ingramcontent.com/pod-product-compliance
Lightning Source LLC
Chambersburg PA
CBHW021508210326

41599CB00012B/1180